Concept Car Design
Driving The Dream

Jonathan Bell

Book designed by dixonbaxi | London
T +44 [0] 20 7864 9993

Production and separations by
ProVision Pte. Ltd. in Singapore
Printed in China

T +656 334 7720
F +656 334 7721

Concept Car Design
Driving The Dream

Jonathan Bell

A RotoVision Book
Published and distributed by
RotoVision SA
Route Suisse 9
CH–1295 Mies
Switzerland

RotoVision SA
Sales + Production Office
Sheridan House
112/116 A Western Road
Hove
East Sussex
BN3 1DD
United Kingdom

T +44 [0] 1273 727 268
F +44 [0] 1273 727 269
E sales@rotovision.com
www.rotovision.com

Distributed to the trade in the
United States by F&W
Publications, Inc.
1507 Dana Avenue
Cincinnati, Ohio 45207

Commissioning editor: Chris Foges
Project editor: Kate Shanahan

10 9 8 7 6 5 4 3 2 1
ISBN 2-88046-564-8

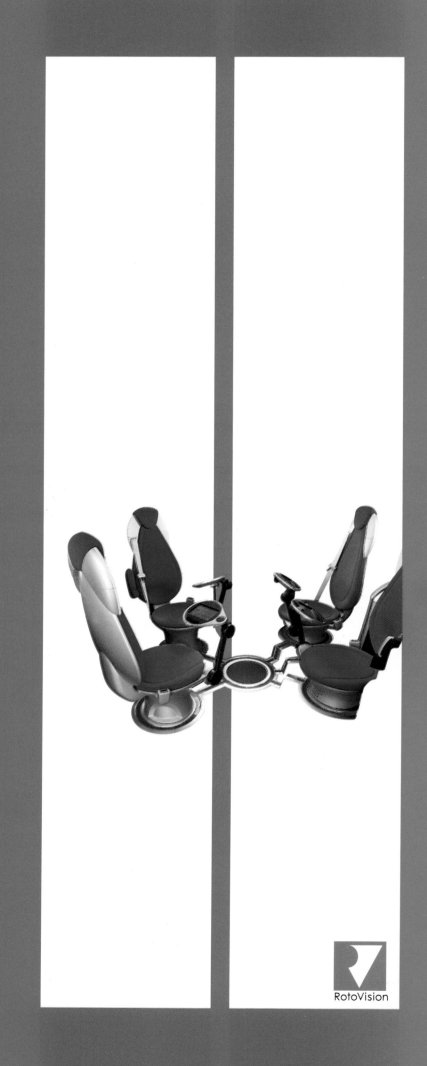

RotoVision

Contents

Introduction

"The concept car is to the production model what haute couture is to ready-to-wear: a reservoir of ideas, a surprising and inviting showcase setting the stage for future production models and reflecting current trends in research, innovation, styling, and creativity."

Patrick le Quément
Senior Vice President, Corporate Design,
Renault

We live in a society that craves the constant rush towards the future. The products that we buy evolve unceasingly, passing on technological improvements, lower prices, and a seductive, ever-changing choice of styles. The automobile industry is no different. We expect car manufacturers to serve us up a diet of novelty, with the old and outdated ruthlessly discarded in favour of new models. Today, it is standard practice for a manufacturer to design and develop perhaps two or three concept vehicles every year, debuting them at motor shows surrounded by media fanfare and maximum publicity. These show cars are, in many ways, the life blood of the industry. They are stylistic test beds that allow the public to see new shapes before committing to the enormous expense of production, and visible indicators of the promise of a better tomorrow, all the better to ensure consumer loyalty. Concepts come from the hugely expensive international design studios maintained around the world by the major companies, worked on by teams of skilled designers with levels of creative freedom quite unlike the restrictive world of the production line.

This is not a book about the history of the motor industry. Nor is it a straightforward catalog of the most famous concept cars. There are many excellent histories of how designers and auto makers have conspired to manufacture dreams, elevating consumer aspirations and sophistication, and driving the market forward in an orgy of futurist optimism. There are also many comprehensive overviews of the industry's more outlandish products[i]. Instead, this book looks at the way in which the concept car has developed, along with its role in the current industry, and the methods – technical, marketing and otherwise – that lie behind its creation. It also looks at how car designers and car companies think, and how the changing social perception of the automobile, our requirements and technological development have informed the cars that come out of styling studios and ultimately the cars that we buy. The gap between fantasy and reality that was embodied by the dream car is closing fast.

In the six decades since the first concept was created, the fundamental principle behind its development hasn't changed. The concept car is the future, a vehicle designed to project a vision of the future, whether that future be months, years, or even decades away. Invariably hand-built, in stark comparison to the production car it imitates, the concept car is a calculated exercise in making the unknown visible, extrapolated from available knowledge, a sneak preview of next season. Generally speaking, the public perception of the concept car stems from a lack of understanding of the motor industry's complex timelines. Today's car designer will already have signed off the replacement vehicle for your current model, committing the manufacturer to every last detail of a product

1958 "Damsels of Design":

A 1958 press shot of General Motors
Head Designer Harley Earl with the
so-called "Damsels of Design."

that won't debut for perhaps two or three years. While this design
is prepared for production – an immensely complicated process –
the drawing boards and CAD stations are already humming with
its succeeding model, two whole generations down the line. For this
reason, the vehicles we see at motor shows, however remarkable
and radical they might appear, are invariably far closer to production
reality than we imagine.

Add this to the immense secrecy that surrounds the industry, and
the concept's prominence begins to look less like a sneak preview
into the industry's inner workings, but more a considered, even
calculated approach. According to the US Commerce Department,
American consumer spending on new cars was $101bn in 1999.
The use of a concept model – costing perhaps several millions of
dollars – is small fry, particularly if it helps gauge public response to
new forms and styles. Traditionally, production technology dictated the
pace of the motor industry. Design, testing, tooling, and marketing a
new model costs billions of dollars – a risky process that, even today,
leaves very little room for error. But as technology reduces the time it
takes to get from design through to development and ultimately the
production line, the role of the concept car has changed.

In the early days of the motor industry's post-war boom, the concept
car was a different beast. (In January 1954, nearly 31,000 people
were crowded into the Waldorf Astoria's Grand Ballroom, New York,
for the opening of the latest General Motor's Motorama exhibition.)
General Motors used the Motorama exhibition to feed the automotive
dreams of the American public, upping the ante and levels of
showmanship year upon year. The first Motorama was held in 1949,
as the depressed post-war car market – thanks to limited production
capability – finally caught up with demand. The extravagant show,
which visited New York and Boston, was seen by nearly 600,000
people, eager for novelty and hungry for the next big thing. By the
early 1950s, the major draw at the Motoramas were the dream cars,
styling exercises that extrapolated the forms available in the current
marketplace, pushing the limits of public taste.

The average all-American family thrilled to the displays, which
ranged from shiny new production cars to kitchen appliances. In the
mid-1950s, design and technology had become the strongest selling
points for new cars, as post-war austerity started to fade and
considerations such as fuel economy and purchase cost were
superseded by horsepower counts, interior space, and luxurious
extras. Our typical family would have spent the preceding years
being tempted by cars like the 1951 Kaiser – "the only car with
Anatomic Design," the 1952 Lincoln, which "surrounds you with glass
– 3,271 square inches… the one fine car deliberately designed for
modern living," or the 1954 Buick Roadmaster, "a car of stature and

1954 Buick Wildcat:

The inspiration for the Chevrolet Corvette, the fibreglass-bodied Wildcat was billed as "The only sports car with truly American styling."

distinction." Styling was the industry's key selling factor, a factor decried only by consumer advocates[ii] and hailed by government and public alike as a economic and social miracle[iii].

While the motor industry was born of engineering, it was the application of style – initially in the form of hand built coach work on standard, mass-produced chassis – that spawned the contemporary car market. General Motors' genius, swiftly adopted by the rest of the industry, was to realize that although technological innovation drove sales, novelty and modernity could equally be implied through the color, shape, and style of car bodywork, in much the same way as the fashion industry generated a constant parade of new lines and fashions. By promoting style as the primary indicator of sophistication, and by incrementally adjusting and developing car styling, the market could be encouraged to buy more, ultimately even changing their models on an annual basis to keep abreast with the latest trend. Design had become big business, and the realization that consumer demand should be sated at all costs convinced the motor manufacturers to emphasize previously unconsidered factors. Harley Earl, GM's Head Designer, introduced female designers – the so-called "Damsels of Design" – to develop cars that might appeal to the female consumer. Initially focusing on interiors and trim ("always on the look-out for anything that might snag their nylons"), Earl noted that, "so many talented girls are entering our field of design, that in

three or four years, women may be designing car exteriors."[iv] Behind these line-ups of fresh chrome that filled the auto dealerships were the dream cars, well-publicized teasers that were seen at shows such as the Motoramas, ogled by an eager public, who transferred their desire into sales. Initially, the dream cars were standard production cars draped in extras, outrageous trims, custom paint jobs, or unusual options. Gradually, dream cars began to evolve into something more. By 1953, GM were advertising "Dreams – on wheels." five mock-up production cars that were stars of the early Motorama shows and illustrated "new processes, new materials, and new techniques now being developed." 1953 saw the debut of Chevrolet's Corvette, a fibreglass-bodied vehicle that was billed as the first American sports car. The Le Mans and Starfire also debuted during this year. Led by manufacturers like GM (see page 38), show cars evolved into highly sophisticated and elaborate one-offs, vehicles that demonstrated the technological optimism of the era – computer-controlled, self-driving, turbine, and even (potentially) nuclear-powered.

The Motorama was a musical spectacular, with shows hosted by models, soundtracked by a full orchestra, with displays centered on the dream cars of the future, such as the Cadillac El Camino, Pontiac Strato Street, and Buick Wildcat. That same year, GM showed the XP-21 Firebird, a gas turbine-powered vehicle that proved a stunning

(From left to right) 1954 XP-21 Firebird, 1956 Firebird II, 1959 Firebird III:

The first Firebird concept car, the joystick-controlled, gas turbine-powered XP-21 was allegedly based on the F-4-D1 Douglas Skyray fighter plane. Features included airbrakes and a 110-volt output for domestic appliances in the event of a power-cut.

Accompanied by the vision of a "safety electronic highway of tomorrow," the Firebird II debuted at the 1956 Motorama. A "blueprint come true," the four-seater II built on the performance of the first Firebird and added technological innovations such as the titanium body, elevated luggage compartment and retractable headlights.

"Imagination in motion," the Firebird III included a working version of GM's 'automatic guidance' system, which used control wires embedded in the road surface. As well as joystick control and remote locking, the Firebird III was "drive-by-wire" – the "first completely electronically-controlled car."

1956 Buick Centurion:

An all-glass canopy roof dominated the Centurion's appearance. The dream car was one of the first to feature a rear-view television camera. This is a perennial concept car favorite, but a feature that has only recently reached mass production in the 2002 Nissan Primera.

(if technically impractical) evocation of Detroit's abilities. 1954 was also the year of the first Ford Thunderbird. By the following year, the Motorama was deemed big enough to be given its own Bob Hope-hosted TV special, with even more elaborate sets and more automotive marvels, including the L'Universelle, a "dream truck" billed as a futuristic combination of van and passenger car.

Subtitled the Highway of Tomorrow, the 1956 show introduced the Firebird II, and five other prototypes – the Chevrolet Impala, Oldsmobile Golden Rocket, Cadillac Brougham Town Car, Buick Centurion, and the Pontiac Club de Mer. Lavishly named and extravagantly appointed, these were vehicles designed to inspire confidence in GM and optimism for the future, as well as bolstering current sales, promising advances that lurked just around the corner. Created under Ned Nickles, head designer at Buick, the Centurion replaced the traditional dashboard with futuristic dials and discs, with a TV camera instead of rear-view mirrors. A huge, clear-glass canopy drew comparisons with the most visible futuristic symbol of the day, the jet fighter, and a "new visual vocabulary, inspired by the jet and the rocket"[v] became the standard sight on American roads. Dream cars were hailed as crucial experiments in future transport: "for many years it has been General Motors' practice to take design and engineering ideas beyond the workshop stage and build them into actual running cars – 'experimental laboratories on wheels.'"[vi]

The 1963 Buick Riviera Silver Arrow:

A luxurious concept proposal from Buick that evolved into the
Riviera production car.

For the most part, the dream cars of the 1950s and 1960s remained
just that, fantasies with features and demands that exceeded practical
application. Detroit rode on a wave of high popular esteem –
practically anything seemed possible. The Firebird III, shown in 1959,
completed the company's gas turbine triptych, and was perhaps
the most radical concept ever shown to a mass audience. Along with
advertising, GM's Motoramas played a crucial part in promoting the
annual model change, and thus the industry's drive for profits.
Although Harley Earl's celebrated Y-Job (see page 41) predated them
by several years, these shows were the true origin of the concept car
as we understand it today – a unique styling statement intended as a
showcase for forthcoming design and technology.

Today, single manufacturer road shows have been dispensed with in
favour of the global motor show circuit and an established system of
press launches with the concept car at its heart. Jump forward to the
Cobo exhibition hall in downtown Detroit, or Tokyo's Nippon Design
Center and the scenario hasn't changed enormously in 50 years: the
system is structured along similar lines. The major manufacturers
who make up GM's global portfolio, for example, are still developing
dream cars for an eager public, only "dream cars" are now concept
cars, trailed by teasing press releases, with glossy information packs
and stands designed by leading design consultancies, topped off
with theatrical unveiling ceremonies.

The term "concept car" was reputedly coined in around 1980,
turning these cars from unachievable dreams into more tangible,
realizable objects. Concepts are no longer just projections of the
future – they also act as cultural triggers, deliberately evoking the
past, unlocking and reawakening subconscious memories and
desires. Concepts are also the proving ground for new types of car,
fracturing the industry into hundreds of market segments. The advent
of highly specialized market research meant that manufacturers
could accurately predict who would want to buy what, and, for the
first time, flexible mass production enabled products to be
specifically targeted at these identifiable lifestyle groups. Although
the motor industry was relatively inflexible in relation to, say, a fashion
house, the 1970s and 1980s saw the concept become a search for a
new type, not just style of vehicle. The "any color you like as long as
it's black" mentality had faded swiftly into memory with the advent of
the dream cars, but the options and extras that proliferated were
accretions, things that made your car feel "fully loaded," but not
fundamentally different from your neighbor's.

Consumer taste has become more sophisticated. Concepts are no
longer just carrots dangled at motor shows to encourage sales of
production-line vehicles. Today, car styling has to appeal to an
increasingly expanding and ever-more complex set of consumer
personalities. The word "lifestyle" – coined to describe the myriad

1956 Oldsmobile Golden Rocket:

The Golden Rocket had a trimaran-like body, with
the radiator protruding out in front of the two wings. Opening the
door set off a series of motorized actions to rotate the seat and
raise the roof, all for ease of access.

consumer categories that arose out of the new self-aware society –
dictates the way in which products are designed and sold. Cars are
no exception. The car is perhaps the most visible of any consumer
product, a mobile symbol of status whose ubiquity and multiple
layers of meaning and value functions as highly effective cultural
communicator; cars act as vehicles for our personalities and
aspirations, not just as means of transportation. These shifts in the
way cars are presented and sold have been achieved with the
development of new design and manufacturing technologies,
technologies that have brought the radicalism of the concept to the
marketplace. Increasingly, yesterday's concept car is tomorrow's
production reality. The following chapter looks at the evolution of the
concept: the ways in which ideas are conceived, presented and,
ultimately, manufactured.

Footnotes

i   Detroit Dream Cars, John Helig (Motorbooks International, 2001); A Century of Car
     Design, Penny Sparke (Mitchell Beazley, 2002); Concept Cars, Jonathan Wood
     (Parragon Publishing, 1997).
ii   The Waste Makers, Vance Packard (David Mckay Co., 1960) was an early
     investigation into the motor industry's modus operandi, condemning the obsession
     with annual change and novelty as wasteful and inefficient, and above all, a deceit
     practiced upon the general public.
iii Populuxe, Thomas Hine (Bloomsbury, 1987) described the decade from 1954 as one
     of America's great shopping sprees, driven by constant stylistic novelty, in consumer
     goods as well as the cars that came out of Detroit.
iv   General Motors press release (26 May 1957).
v   Populuxe, Thomas Hine p87.
vi   'The Story of Firebird II', General Motors pamphlet (1956).

Chapter One:
Methods

Car design is constantly evolving; a technologically hungry industry that gets immediate, perceptible benefits from new methods.

Car design has a formal, almost classical structure. The various parts of a car bear names that are as fixed and concrete as the elements of a classical column. There is an international vocabulary of styling terms, mixing English and Italian words, creating a design language understood by stylists the world over. A vehicle's beltline, for example, is the zone that divides the greenhouse (window area, or *padiglione* in Italian) from the main body of the car. ''Hardpoints'' means two things: as well as being the design stage that follows on from conceptual sketches, it also refers to the zones on a design that cannot be changed, typically the position of certain elements such as window pillars. Just as classicism provided an order within which architects could compose a vast array of building types, so the car's essential components can be rearranged and reinterpreted, the difference being that technology is continually re-defining how materials can be made to perform.

Of course, car designers are at the mercy of engineers, for the mechanical underpinning of a car remains the true structure that must be adhered to. As part of the team developing a design, those responsible for the actual physical appearance of a vehicle are usually described as stylists – a throwback to the traditional separation between a vehicle's skin and skeleton. Despite the popular perception of car design as something that relies entirely on computers, the actual methods of modeling concept vehicles have remained relatively unchanged for 50 years: many concepts shown at motor shows are simply non-functioning fibreglass shells, without opening doors, windows, or interiors. On very rare occasions, the concept unveiled to the public has been constructed from painted, trimmed, and polished clay (although this is relatively impractical for transportation and display). To reach the model stage, computers play a vital, high-profile role, but raw analogue techniques like clay modeling co-exist with slick digital processes, silicon complementing clay and vice versa.

Concept car design is not car design at its purest – however elegant, no engineer would concur that freehand sketches represent the pinnacle of automotive art. Yet free from the considerations that might constrain artistic expression, concept cars allow for the creation of forms that might otherwise never see the light of day. Creating cars specifically for motor shows as opposed to for internal design development purposes also imposes an entirely different discipline, with tighter deadlines and higher pressure. Regardless of the ultimate purpose, the rapid nature of concept car design generates working methodologies that ultimately benefit and improve the integration between the design and manufacturing processes.

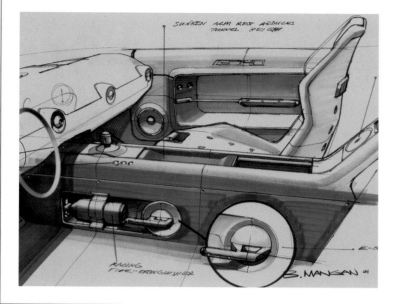

Ford GT-40:

This design sketch explores a possible
interior concept.

New Mini sketch:

Frank Stephenson's stylized sketches link
the new car to its heritage.

The Sketch

At the heart of any design is the sketch. Admittedly, there's no such
thing as a clean slate or blank page in car design. When brands,
marques, and models carry such cultural and economic resonance,
not to mention the enormous costs of developing tooling and
mechanical systems, starting from scratch is impossible. But
designers still place huge emphasis – physical and psychological –
on the role of drawing. While some cars famously began life as a
few scribbled lines on a scrap of paper (Sir Alec Issigonis's oft-
reproduced sketch of the fully-formed Mini, for example, slyly
mimicked by BMW's own press imagery), the contemporary
designer's notebook is unlikely to contain such conclusive results.
Instead, the sketch is a means of exploring the shapes and features
that comprise a car and, more importantly, a particular brand of car.
Central to a vehicle's branding is its down-the-road-graphics (DRG) –
the features that define a particular model (the distinctive BMW
radiator, for example). While not set in stone, a marque's DRG is
crucial, a device a stylist interferes with at their peril. The designer's
relationship with marker, pencil, pen, and paper is intuitive and fluid,
allowing for a depth of expression and exploration that even the most
sophisticated computer package cannot match. Scaling up the
finished sketch by translating it into a dedicated 3D package such as
Autostudio is the next stage.

The Model

Car design is industrial design, and while the concept operates at
the rarefied artistic end of what is essentially an industrial process,
the final result must still be translated into three dimensions.
Engineering considerations play a part in every stage of a car
design, even with concepts, and so creating a 3D model is essential.
Translating a drawing from two to three dimensions is at the heart of
the design process. The most intuitive and obvious way to build a
car is to shape it. Motoring lore has it that Harley Earl dug his own
clay from the banks of a California stream, establishing this rough
and ready form of solid modeling as the industry standard. Clay is
pliable and easily shaped, allowing a car design to be a true work
in progress in three dimensions at one-fifth, two-fifths, or full scale.
Typically, a thin layer of modeling clay is supported on armatures
constructed from plywood and foam in the rough shape of the
vehicle. This armature, or "buck," can be mounted on an existing
chassis, with wheel arches created from a more solid material such
as wood or metal. Typically, a styling model might have two subtly
different sides – Janus-like – as a cheap and easy way of exploring
two possible design solutions. The buck is then sealed, ready for the
application of modeling clay. Clay has moved on from the natural,
river bank clay of legend, and companies such as the New Jersey
firm of Chavant manufacture special automotive clay that becomes
soft and tactile when heated yet hardens solid at room temperature.

The Keith E. Crain Transportation Design Center:

This is located in the Walter B. Ford II Building at Detroit's CCE.

[photo: Balthazar Korab, 2001]

Renault Technocentre:

Scaled-down clay bucks are modeled in the Technocentre's design studio.

The clay surface, which is rarely much more than an inch thick, is sealed then painted using standard primer and auto paints, or with Dinoc, a flexible vinyl fabric that replicates the appearance of painted metal. The painted body is then given its "jewelry," the name given to the extra parts such as door handles, lights, flashes of chrome trim, and so on, that make a clay model indistinguishable from the real thing.

Styling a new car therefore becomes akin to sculpture, with experienced modelers demonstrating skills that wouldn't be out of place in the world's great art academies. Computer techniques have evolved to co-exist with this traditional art. Scale models can be swiftly and accurately sculpted by computer-controlled milling machines. These immensely sophisticated, robotically-controlled devices precisely output files from CAD software as physical forms, whether in clay, foam, or synthetic wood. These machines, manufactured by companies like MECOF in Italy, effectively act as three-dimensional printers, outputting forms for appraisal. Naturally, the vast site-specific machinery can receive data from all over the world. The process also works in reverse: clay models can be scanned in three dimensions and turned into computer models. Clay sculpting was traditionally the first three-dimensional scale manifestation of a design, which would have been worked up from sketches and small models. Such was the speed of the computer's

rise to prominence, that many manufacturers believed that clay's days were numbered. The gradual realization that old-fashioned modeling techniques contain a human element that is near impossible to simulate has sent many manufacturers back to the design schools to ensure the craft's survival. The clay buck is also vital for creating a mold for a more hard-wearing fibreglass concept which can be produced in a limited run for display purposes and testing before focus groups; it can also be used for wind-tunnel tests. Today, clay models might only be constructed once a design has been fully developed on a computer, as in the case of the Ford Puma (which was developed in four months). Once the design is agreed, a full-size clay model is made for initial presentations. Clay is heavy and not especially hard-wearing. A completed model might weigh as much as four tons, with the clay layered from anything between one inch to a foot on the surface of the buck. A forklift is needed to move the model around the studio, so the fibreglass molds are essential if the concept needs to be moved efficiently and safely.

Design sketches for the Renault Talisman concept:

The working sketch on the far right shows the four individual instrument pods attached to the steering column, which made it into the final concept in a slightly modified form.

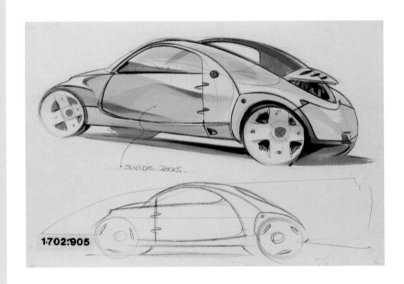

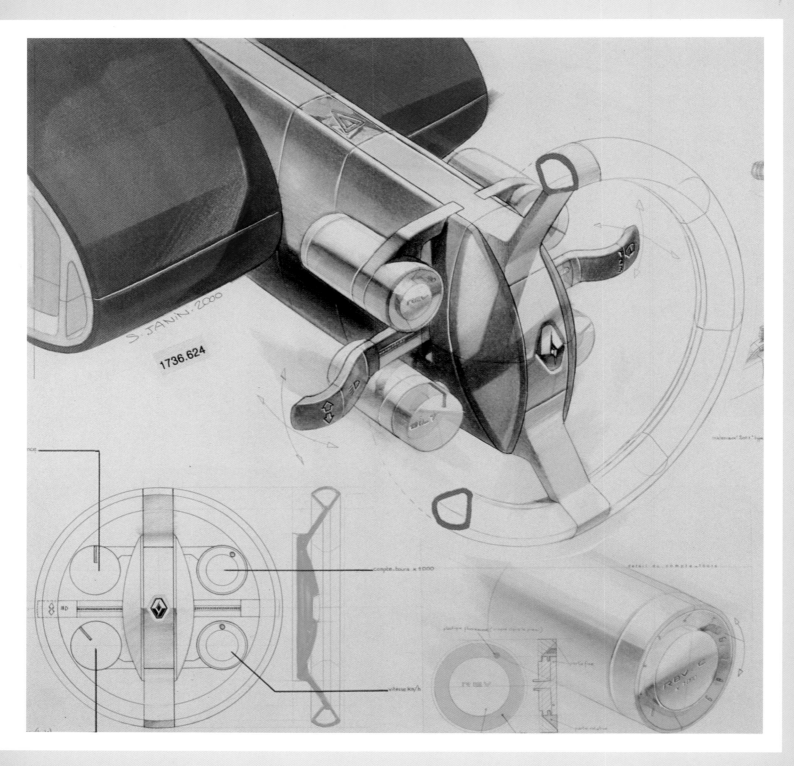

S. JANIN. 2000

1736.624

Modeling a clay buck to scale:

Producing a full-sized clay model of the GT40 concept, using hand sculpting techniques in conjunction with a Tarus computer-controlled milling machine.

exterior color
blue

TPI TARUS

Chrysler:

Virtual representations of Chrysler
concept vehicles, rendered in the
industry-standard CATIA design software.

The Technology

The technology that facilitates the transition from paper to model
has undergone a revolution in the past three decades. Few CAD
packages offer comprehensive integration of all computer-assisted
processes – styling, modeling, component stress analysis,
aerodynamic testing, etc. However, the industry standards, such as
Alias's Autostudio, Computer-Aided Three-Dimensional Interactive
Application (CATIA) and IDEAS, can share data between each other
using a wide variety of engineering software. Alias drawings can be
modified with the package's Paint section, then 3D data is input into
a computer milling machine to produce swift scale models. A virtual
model can be crashed against a virtual wall (courtesy of PAM-
CRASH from France's ESI Group), driven around an imaginary
test track (using MSC Software's ADAMS program), placed in a
virtual wind-tunnel using computational fluid dynamics software
and even built in a virtual factory. Computers have cut development
cycles from five years to three, and concepts are produced even
more swiftly.

CAD

By streamlining their design systems with a single CAD package
that is globally adopted by all studios, a manufacturer can ensure that
rolling concepts can be treated almost like sketches. Thanks to the
use of a data warehouse (holding some 340 terabytes of digital design
data) GM, for example, shares information between all its design
studios, vastly speeding up design time. From 2000, the global giant
introduced a new model nearly every month for two years (even
designing a concept, the Buick Bengal, around a celebrity – in this
case, the golfer Tiger Woods). The heady speed of these development
programs has completely superseded the motor industry's traditional
reliance on one annual forum for new models (usually the most
prestigious motor show in the manufacturer's company of origin).
Instead, new cars are expected throughout the year.

CAD has developed alongside the auto industry. In 1963, the first CAD
system was delivered to General Motors. DAC-1, standing for Design
Augmented by Computers, was developed by IBM and Don Hart and
Ed Jacks of GM's research team from the late 1950s onwards, which
coincided with the debut of Ivan Sutherland's Sketchpad program.
Sutherland started work on Sketchpad in 1960 as part of his PhD
researching computer graphics and engineering design, developing
a system that not only allowed operators to draw on the screen with a
lightpen, but also permitted individual objects to be stored and

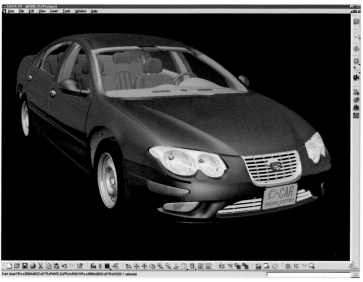

manipulated. A simple 3D model of a design could be specified and then rotated and viewed from different angles. GM was also using digital and analogue computers during the 1950s to help solve engineering problems. Development of these early CAD systems was accelerated by their application in the engineering and aerospace industries, and car manufacturers benefited from bumper Cold War budgets.

CAD, together with computer-aided engineering and manufacturing (CAE and CAM), improved in quality as computing power increased exponentially. Software innovations like solid modeling were introduced throughout the 1980s, allowing a virtual sense of a material's form, properties, and visual appearance to be ever more accurately gauged on a computer screen. Despite the great advances in computer visualization, only by using a full-size, painted model can designers ascertain how factors such as daylight, sunlight, reflections, and shadows play and alter the carefully conceived forms of a car. A major design studio will typically have a daylight viewing area – a carefully concealed courtyard where models can be assessed without revealing them to outside sources. "We're not comfortable designing completely in the computer simply because you don't have the ability to walk around the car in real time." J Mays admitted in the late 1990s. "We simply don't have a big enough screen to accurately portray the size of something like an automobile."[1]

Visualization Systems

Manufacturers now possess visualization systems that come close to the qualities of clay models. DaimlerChrysler's Virtual Reality Center is located at the Mercedes plant in Sindelfingen, Germany. Fully integrated virtual reality systems are becoming an essential part of the design and development process, allowing potential problems to be tackled earlier. The company makes use of three main virtual presentation methods; the Powerwall, the Cave, and a cylindrical projection area which allows for the panoramic projection of models. The Powerwall is 23 feet by eight feet, a huge screen onto which design studies can be projected, considered, and ultimately signed off for the next stage of development. Powered by a group of high-end Silicon Graphics Onyx2 InfiniteReality2 systems, the Powerwall can render virtual crash tests, as well as make useful comparisons between rival models. The Cave offers more interaction. With projections on the walls, floor, and ceiling, it provides an immersive virtual environment that can be used to assess ergonomics, color schemes, instrumentation, and so on. When designing production cars, the moment when tweaking stops and a design is committed to is known as the design freeze. Once it has been imposed, three pre-production models are prepared, each demonstrating increasing levels of production-ready technology, building up to the actual launch of the production line and hence the new model.

Methods

2001 Buick Bengal:

A new direction for concept marketing, the 2001 Bengal Concept was initially launched in association with golfer Tiger Woods. Apart from the weak pun on the name, the car, which might see production in a modified form, was apparently designed with extra space for golf bags.

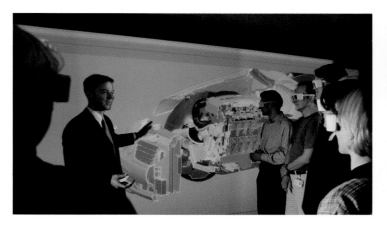

DaimlerChrysler Virtual Reality Center in Sindelfingen, Germany:

Opened in 2000, DaimlerChrysler hoped the high tech visualization center (users are seen discussing three-dimensional data on the 23 feet x 8 feet Powerwall display) would speed up prototype development times.

DaimlerChrysler Virtual Reality Center in Sindelfingen, Germany:

The Cave is a five-sided space for projecting car interiors. Each side of the acrylic glass structure is powered by its own SiliconGraphics workstation.

Concept Limitations

Concepts are rarely pre-production models, and the emphasis on new technology, materials, and untried shapes means that the average concept is unlikely even to be self-propelled. Show concepts, or platform styling models, might not even have real turning wheels (and have plastic ones instead). Interiors – if a whole interior is actually built – are usually furnished with dummy instruments. Some styling concepts might not have seats or trim, just a dark cloth panel at dashboard height.

Education

At industry level, the integration of the design process has seen an acceleration in the time it takes to develop and manufacture a rolling concept. The mechanisms and systems that lie at the heart of car design have been subject to constant flux and development, and keeping in touch and on top of new technology is essential. This has important implications for design education. As well as being taught how to draw and make models using traditional processes, car design students are faced with a dizzying array of computer packages, all of which promise to mimic and improve upon the human hand and eye. There is no one industry standard software package, or, for that matter, any standard means of integrating analogue and digital processes.

Design schools such as Pasadena's Art Center College of Design in the US (which has produced approximately half the world's car designers), Hongik University in Korea, Strate in France, and Coventry University and the Royal College of Art (RCA) in the UK, amongst others, provide practical training and instruction on the tools of the trade. Sponsorship and collaborations with the industry go some way to replicating the conditions in a professional design studio. The RCA established the first vehicle design course in 1967 and to date it can claim that over 150 vehicles have been designed by its graduates, including the Lotus Elise, McLaren F1 and Audi TT[ii]. For decades, the most talented students from Pasadena, the RCA, and Coventry have left their respective courses and stepped straight into the studios of the major manufacturers. Despite this, students are able to infuse their work with a more personal agenda, encompassing radical, political, and social elements that might not find expression in a big auto studio. The emphasis in design schools is not on pure automotive design, but also on other transportation projects, be they small-scale, lightweight proposals for motorbikes, bicycles, yachts, catamarans, and even Zeppelins, or public transport such as buses and trams.

Student concept vechicle design:

Adam Hopwood's concept 'Sunraycer' was designed to use solar power. Hopwood graduated in 2001 from Coventry University in the UK with an MA in Automotive Design.

Student concept vechicle design:

Jorge Diez worked on modeling from RCA concepts before graduating in 2001 from the RCA in London.

[Photo by Eileen Perrier]

Sponsorship in Education

Sponsorship is crucial. Nissan, for example, sponsored the 2002 Vehicle Design Competition at the RCA, a post-graduate module looking at how a compact Nissan car might appear in 2002. In particular, students examined how a car might develop with age, and how the changing demographics of the car buyer would be reflected in its design, as well as investigating new forms of visual language for the Nissan brand. Coursework included interviewing car buyers and even "used new radical research methods such as visiting fortune tellers to predict future trends."[iii] Student work of this sort is valuable to the companies that mentor and sponsor it, as the emphasis placed on relating to the meaning, values, and future of a brand is essentially free creative research. Coventry's 2001 students were asked to look even further into the future and create a car for 2030. Sponsored by steel maker Corus (which has a large automotive division), the students had to convey an understanding of developing manufacturing, technological, and material demands, as well as look at styling. Barcelona's Elisava School of Design has collaborated with Airbus and SEAT: link-ups between education and industry are too numerous to mention. The Transportation Design Forum, now in its second year, is held at Pforzheim University in Stuttgart, which has an established transport design course. This provides an overview of automobile design education, bringing together the work of students worldwide to be assesed by established industry designers.

The Student Experience

Despite the sponsorship opportunities and funds given to car design courses by car manufacturers, student facilities are understandably less comprehensive than those found in industry. Students therefore spend the majority of their time concentrating on two areas: 2D sketching (both by hand and digitally) and the creation of scale models, either in clay or other modeling material, or by using a computer (small-scale rapid prototyping facilities are also becoming more widespread). For the design student everything is a concept, a leap into the void, a personal exploration, regardless of the rigidity of the brief, or of a model, marque, or genre. Once a student enters a real design studio, the reality is more mundane. All the 10,000+ components that make up a contemporary vehicle have to be designed, meaning long hours slaving over CAD models of door knobs, internal lights, and a thousand other vital details. There is a world of difference between developing all the elements of a single, stand-alone model and working in a real studio. The emphasis shifts to the micro level – the larger the studio, the more likely it is that a new designer will start off working on small items of trim – such as door-handles, wing-mirrors, or hub-caps. Working in a dedicated concept studio is exciting, but rare.

RCA:

The emphasis in the design schools
is not purely on automotive design.
This three-wheeled mode of transport
was proposed by RCA graduate
Bruce Wheatley.

[photo: Dominic Sweeny]

Concept Origins

Concept cars remain the industry's focal point, the media-friendly products of design studios that justify the design department's prominent role in the maintenance, promotion, and development of a company's brand image. Nevertheless, the cost to a manufacturer of maintaining an international network of design studios is vast – figures of up to $100m a year have been quoted.

Originally, dedicated studios were the preserve of independent companies, where designers, stylists, and coachbuilders were contracted by the big auto giants to clad their mechanicals in sumptuous bodywork. Having evolved from the earliest days of the auto industry, the coachbuilders (or *carrozzeria*) were in a position of great influence, even as the introduction of unitary construction (whereby the chassis and body are one structural entity) put paid to their original role. Coachbuilders were the industry's style specialists, in great demand for limited editions, new models and, perhaps above all, concept cars.

But gradually the great names in coachbuilding slipped away, all too often amalgamated into the companies they once supplied, their long and prestigious histories reduced to a mere badge on the bonnet of a high-specification model. Ghia, acquired by Ford in 1973, was recently downsized from a coachbuilder, which had been complete with the tools and equipment needed to make full-size concepts, to an "electronic concept studio," similar to that run by Volvo in Barcelona. Other companies, such as Pininfarina, founded in 1930 and best known for collaborations with Alfa Romeo, Ferrari, Fiat, Lancia, Peugeot, Bertone and Zagato, have been able to retain a high degree of independence, evolving into highly-skilled but relatively low-volume manufacturers, like Bertone and Ghia. The German company Karmann builds just 100,000 vehicles a year, and has collaborated with the likes of Volkswagen. They will build the new Chrysler Crossfire, a subdued version of an earlier concept.

Current Trends

The technologies listed earlier in the chapter have splintered the industry's traditional categories, sub-dividing the car market with more models, each aimed at different consumers. Independent designer Ken Okuyama, with experience at GM and Pininfarina and the man behind Pininfarina's Metrocubo, a city vehicle described as the iMac of the auto industry, believes that, "niche vehicles will be the dominant trend in the future."[IV] Like many other commentators, Okuyama foresees an end to the mega-trends that shaped the industry for so long – angular styling, aerodynamic styling, city cars, SUVs, etc. "Before," he says, "you could determine whether the future of car design was say, either round or square, but now you can't tell, those waves don't exist anymore."

RCA:

Motorcycle design by
RCA graduate Kristain Hardy.

[photo: Dominic Sweeny]

The fact that the great majority of concepts are not intended for production means that concept design avoids a great deal of the logistical complexity inherent in designing an all-new model. However, as discussed in the introduction, the increasing cross-over between conceptual and production is massively assisted by the integration of the various systems – design, prototyping, engineering, tooling, and testing – within a modern car company. The sheer speed of today's system is best illustrated by a recent project undertaken by the British magazine Autocar and niche engineering experts TWR. Under the title Project 3PV, the team attempted to go from concept to reality in a mere 15 weeks. Briefed to design a small second car for an average-sized family, TVR designed a three-seater, producing eight initial concepts in sketch form to explore internal layouts and body shapes. Once a design had been chosen, detailed studies were made, drawn up on computer in order for engineering absolutes – drivetrain, powerplant, cabin dimensions, etc. – to be plotted and accommodated. Alias drawings were then plotted into a computer-controlled milling machine and output first as a scale clay model, then as a full-size mock-up. While the finished concept was not slated for actual production, the exercise was a good demonstration of the speed of today's industry.

Mainstream manufacturers are honing their design processes. Renault manage vehicle design using a Product Planning Department, one which "carries out detailed socio-cultural research in conjunction with groups that specialize in population studies, social development and consumer behaviour." [iv] The ambitions for a new car are defined in a vision plan, a short document outlining a manufacturer's intentions. This brief generates feedback from the design, manufacturing, and marketing departments, all three of which work simultaneously on a particular project. While design might progress from mood boards and sketches, through to clay interpretations of the proposed design, engineering departments need to ensure that design requirements are accommodated by more fundamental concerns. These include the layout of the manufacturing plant, availability of materials, development and design of new tooling, not to mention the development and testing of new components. Design and manufacturing information is typically stored on a shared database, allowing easy communication between dispersed sites.

Some firms use a central site for design, such as Renault's Technocentre in Guyancourt, Paris. The Technocentre, established in 1998, houses facilities for 7,500 engineers and technicians, and was one of the first locations for a real-time, life-size 3D virtual modeling system. This is similar to those maintained by DaimlerChrysler in Germany, TWR's Leafield Reality Center, and Worthing Studios in the UK (where the 3PV was developed). Renault were pioneers in

2001 Chrysler Crossfire:

An American take on the TT, the
Crossfire will enter production in 2003.

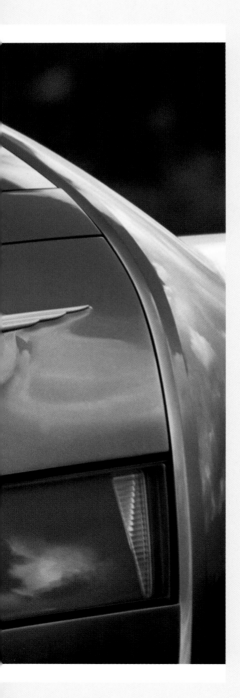

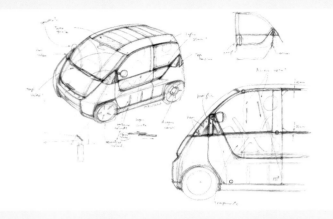

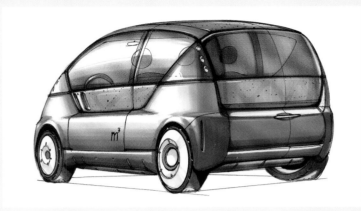

Pininfarina Metrocubo:

A tiny city car, the Metrocubo had none of
Pininfarina's traditional flair, but was eminently practical.

computer techniques; for over a decade the design department has produced short films that blend computer-generated concepts with live action back environments, where whole cities are populated with virtual representations of the cars of the future.

Car design is constantly evolving; a technologically hungry industry that gets immediate, perceptible benefits from new methods. By ensuring that the systems are in place for future growth, together with a strong education network and the new breed of dedicated design studios, manufacturers are committed to exploring the further reaches of car design.

Footnotes

i   www.thing.net/~lilyvac/mays.html, J Mays in conversation with artist Dike Blair

ii   Moving Objects, the catalog of the RCA's 30th anniversary show (2000), contains a comprehensive list of past students

iii   Nissan press release (19 February 2002)

iv   Automotive Design and Production magazine, Ken Okuyama, interviewed by Kermit Whitfield (2 January 2002)

v   www.renault.com website (May 2002)

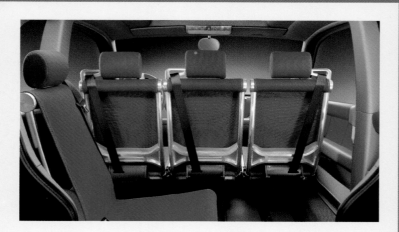

Industry behemoth:
General Motors

The GM Technical Center:

Opened in 1950, the General Motors Technical Center was designed by Eero Saarinen. The 330-acre complex, centered on a 22-acre lake, includes machine rooms, laboratories, and styling studios. The Center is located to the north of Detroit, USA.

Industry behemoth:
General Motors

The term "concept car" might be a recent invention, but it owes a great deal to the work of General Motors in the 1930s. This behemoth amongst car manufacturers shaped the modern industry, from its role in the creation of the car dealership system in the late 19th century, to the way in which it encouraged the shift from public to private transport after the Second World War. GM's Art and Color Division was arguably the industry's first styling studio. Headed by Harley Earl (1893–1969), the son of a coachbuilder, who specialized in creating lavish custom models for the Hollywood elite, GM popularized the dream car, using motor shows and special exhibitions to display innovative and eccentric styling ideas to a mass audience. Through the celebrated post-war Motoramas, culminating in the Futurama display at the 1964 New York World's Fair, GM laid down the first template for the use of concept cars: the future made real.

Art and Color was established in 1927, under GM's President Albert Sloan. The department's name indicated the industry's initial simplistic approach to design and marketing. In 1937, Earl renamed it the Style Section, and produced the first physical manifestation of

The LeSabre concept:

Produced by Harley Earl's studio in 1951, this stylish, two-seater evocation of the nascent jet age
was a foretaste of the decade's later excesses. Earl used the LeSabre as his own personal car for
many years.

the design techniques he pioneered, including the use of clay
models. The 1938 Buick Y-Job was a remarkable vehicle, and is
now held in motoring lore as the first true concept. Based around
a production Buick Century, the Y-Job was a vast, 17-feet, two-seat
convertible. Earl instigated and oversaw the project, the concept
was styled by George Snyder and engineered by Charlie Chayne.
Stylistically, the absence of running boards and integrated front wings
predicted the post-war aesthetic, and other innovations included
pop-up headlights. Most importantly, the Y-Job owed its existence as
much to canny marketing as to a genuine desire for technological
improvement: Earl named the car "Y" as this was one stage on from
"X," the traditional engineering designation for experimental.
Earl drove the Y-Job, but favored GM's first major post-war concept,
the 1951 Le Sabre, as his personal day-to-day transport.

After the war, GM opened the Technical Center, an advanced
research facility where vehicles could be rigorously tested.
The Technical Center provided engineering back-up for the
concepts most notably for the famous Firebird series, which

incorporated technologies such as a gas turbine powerplant and
"Autoglide" self-steering. The 1958 Firebird III also featured one of
the first cruise control systems as well as the "Unicontrol" steering
system, a single joystick control (a feature that has never made it to
a production model, although Mercedes' gull-winged 1996 F200
concept and several other concepts have adopted a similar system).
By the time Earl retired in 1958, the company's design and styling
department had a staff of over 1,000.

The electronic studio:
IDEA

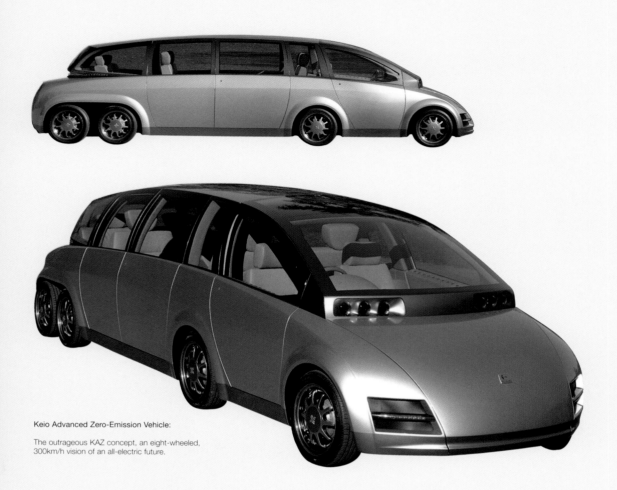

Keio Advanced Zero-Emission Vehicle:

The outrageous KAZ concept, an eight-wheeled,
300km/h vision of an all-electric future.

Founded by designer Franco Mantegazza in 1978, the Institute of
Development in Automotive Engineering (IDEA) began with work
for Fiat, developing the now-common process of sharing a single
engineering platform across several varied models. Unlike Bertone
or Pininfarina, the Institute has no facilities for production, even on
a limited scale. The studio works on a variety of transportation
problems, including industrial design and the occasional foray into
architecture, but retains an automotive focus. Specializing in melding
the practical with the unusual, the 2001 Keio Advanced Zero-Emission
Vehicle (KAZ) was a concept in the grand, dream car tradition of the
word. Developed in collaboration with Japan's Science and
Technology Corporation, the KAZ is 23 feet long, an eight-wheeled
foray into a fast, all-electric future. A long, extruded glasshouse sits
above the lengthy wheelbase, with four axles (arranged with a pair
of axles at the rear) planned for maximum access to the interior.
Electric motors would be housed within the wheels, and the Institute
claims that speeds of up to 186mph could be reached. The studio
isn't just concerned with wild speculation.

Their Tata Indiva is a compact MPV with simple, unadorned lines
and a friendly "face." Shoehorning seven seats into the one-box
shape, the Indiva retains its compact proportions with a high beltline.

www.idea.institute.it

Three decades of vehicle design innovation

RCA

Sports two-seater, designed by Ted Mannerfelt, 2002 (RCA)

[photo: Dominic Sweeny]

Three decades of vehicle design innovation

RCA

The RCA's course was founded in 1967, with the first students receiving substantial scholarships from Ford UK. Throughout its history, the RCA has supplied the motor industry with designers, establishing a global reputation as the pre-eminent school for car styling with an illustrious role call of alumni, including Stefan Sielaff (Audi), Gerry McGovern (Lincoln), Simon Cox (GM, Isuzu), Ian Callum (Jaguar), and the late Geoff Gardiner (Jaguar). Course Director Dale Harrow stresses that the RCA places an emphasis on "designing through making." In recent years, the emphasis has shifted slightly from creating concepts that represent an evolution of current styles to a more complex process of synthesizing a particular manufacturer's brand image in order to specifically target a particular market. A typical early brief might be to take a word and then turn that into a concept, which could be described as styling meets semantics. "It's about how you connect with an object emotionally," says Harrow, explaining how students are now interpreters of cultural values. For a case study on Renault, for example, students would study all aspects of French culture, attempting to imbibe their concept with a Gallic essence.

A futuristic sports coupé:

This coupé was badged as a Cadillac, and designed by
RCA graduate Andrew Sheffield.

[photo: Dominic Sweeny]

Harrow acknowledges there has been a shift away from skills building.
Today's students will be facing different issues in their career from
that of a decade ago, as computer technology redefines the ways in
which concepts are created. "We're trying to get students to find out
the big issues and bring that to their work," says Harrow, explaining the
importance of a global overview. Students are now expected to be
sharply attuned to all aspects of contemporary culture, not just well
versed in the myriad histories and names that every petrolhead
learns from an early age. The RCA's current approach seems emotive
– a marked change from the original role of "styling," seen as mere
dressing for engineering, derided by automotive engineers and
berated and sidelined by design commentators.

Car design is a long-term game, and theory and research play an
increasingly large role. Partly this is catered for by PhD research, and
more focused elements such as the RCA's Helen Hamyn Research
Centre for inclusive design. Harrow also takes the long view that the
car market will change significantly in the next generation, although
the raw emotion of car design will be hard to lose.

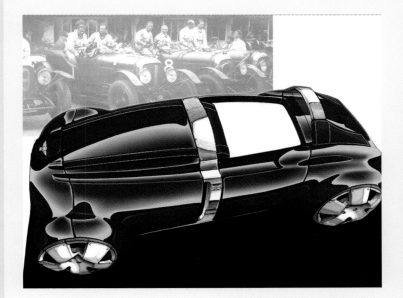

Ford, Bentley and Mercedes concepts:

Final year sketches by Coventry Transport Design students Gareth Thomas, Jay Lee and Dimitrios Zacharopoulos, 2002. The Ford-badged concept on the far right is described as "a vehicle that can satisfy the needs of the urbanite."

NISSAN GT-R:

A star of the 2001 Tokyo Motor Show, the GT-R concept updated the image of the company's high performance GT-R brand. Simple, almost brutal styling highlights the car's technical and sporting ability.

Industry and education
CCS-Michelin Design Competition

-wind reflector
offers an open
road feeling

-prominant rear-
view mirrors

-machine like
aesthetics and
exposed parts

CSS Design Competition entry designs (above and left):

Michelin concept tyres (right):

Mitsubishi collaborated with Michelin to develop a tyre suitable for dry traction and
maximum urban handling. The blue enhances the visibility and acts as a styling cue. The
other tyre was developed with General Motors and features a unique wraparound tread
design to enhance a vehicle's profiled appearance.

As one might expect, student entrants into the annual Michelin
College for Creative Studies (CCS) Design Competition focus on
the sponsor's main assocation with the auto industry: the tire.
Held for 13 years in collaboration with Detroit's CCS, the design
challenge encourages students to apply real-world technologies to
concept designs, as well as looking at ways in which contemporary
technologies might be improved or adapted to suit future vehicles.
Michelin has supplied tires for many concept vehicles, including the
"Inuit" tire for the Bertone Kayak model, the Pax system used on the
Citroen C3 Lumiere – the forerunner of today's production C3, and
the tiny wheels for the diminutive Pininfarina Metrocubo. Student
designs looked at the tire and how it related to the rest of the vehicle,
even incorporating power plants and suspension structures within the
actual structure of the wheel itself, or a tire that could be adapted
instantly from on- to off-road use.

www.ccscad.edu

The design center as publicity machine
Nissan

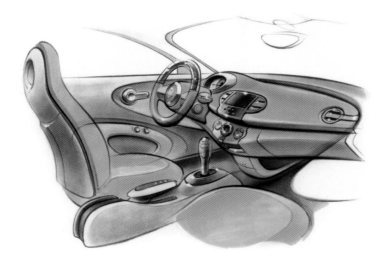

Sketches for the Nissan mm.e concept:

This early depiction of the mm.e includes side spats covering the tyres. The final concept, shown first at Frankfurt 2001, was more conventional, although it kept the high, bonnet-mounted headlights, and gave an indication of the latest generation Micra.

Nissan's design division has a lavish publication for promotional use, illustrating the best recent work from their global Design Centers (Nissan Design have studios in the UK, Germany, San Diego, Michigan, and Japan), as well as containing an extensive section entitled "Showcase of our Heritage." This is the design center as publicity machine. Nissan also sponsors student courses, most notably at the RCA in the UK, illustrating the key role that patronage and sponsorship plays in shaping and directing inspiration. Overseen by various facets of the automotive industry – from car builders to component and material suppliers, student design competitions and sponsored modules provide a detailed brief, right down to the suggested consumer of the end project (thanks to the intense research all car companies carry out).

In March 2002 the company announced that its European Design Center would be located in London, at the Rotunda building in Paddington. "We wanted our designers to be influenced by the major contemporary art, architecture, fashion, and design scenes, which are so vibrant in London," said Shiro Nakamura, the company's Head of Design. Recent Nissan concepts include the Yanya, a compact, robustly styled 4x4 aimed at a young audience; the sporting GT-R; the rugged Crossbo; the mm.e and MOCO compact cars; and the Kino compact minivan.

www.nissan.co.jp/design/

Methods

Motorcity Europe: the independent studio

David Hilton

L-GT Supercar concept:

This full-scale model is the result of collaboration between Ford Design (clay modeling), Bertrandt Gmbh (modeling), Zeltec Gmbh (hard parts) and Motorcity Europe (design). Intended to be a road going GT Le Mans style supercar, it is an inspirational mix of GT-40, Stratos, Batmobile and the Millenium Falcon.

Methods

Motorcity Europe: the independent studio
David Hilton

2001 MONDEO

Concept sketches:

David Hilton's concept sketches for a performance version of the re-styled Ford Mondeo, 2001 (above) and the Ford Focus RS development (left). The RS is a high-performance version of Ford's mass market car. Hilton used Alias Studiopaint to create these images, imagining muscular, strong forms to emphasize the car's power.

Motorcity Europe is a contemporary styling company, owned and operated by designer David Hilton. Working for a variety of major industry players, Hilton is contracted to undertake design, concept, and model-making services. Recent work has included the design of a new supercar, the L-GT, a collaborative project involving the modeling expertise of Ford Design and Bertrandt Gmbh. Hilton typically has a team of between one to 100 designers; the L-GT project involved two designers, two Alias operators and about six modelers. Now based in Cologne, Motorcity Europe was established after an eight-year period at Ford, following Hilton's design, fine art, and product design education at Cincinnati. Hilton has consulted for Mazda, Cadillac, VW, Jaguar, and Ford – his 1992 concept for Jaguar anticipated the eventual shape of the S-Type (perhaps unsurprisingly as the brief was to develop the firm's classic MKII in a modern idiom).

Working methods are a mix of old and new technology. Tools of the trade remain a notebook, pencil, a Wacom pad, and a machine running Alias, the industry standard modeling package. Intitial pen and paper sketches are transferred into Alias, creating a model which can then be re-drawn and sketched over using Alias's 2D tools – "like photo re-touching." Finally the concept is taken into clay.

"Unfortunately in this business I need more technology than the world can offer," Hilton admits, "I need a workstation that fits in my pocket and can do models, renders and sketches." Hilton believes in the future of clay modeling – "there'll always be a need for a full-size clay model to design cars – because it's such a big product you have to see it in full size." He emphasiszes the sculptural quality of concepts, citing the Ghia Focus, sold by Christie's in June 2002 for $1.1m – "illegal on the street, so it was bought as a piece of sculpture."

Consulting designers are relatively rare in the industry, but apart from the obvious cost benefits, it allows manufacturers to bring in completely "fresh" thinking, the same logic behind the global network of semi-independent design studios.

www.motorcityeurope.com

Chapter Two:
Expression

"A concept vehicle should have some dream-like qualities but it shouldn't be so far away from reality that it is just a dream. It should project a future that's reachable."

Simon Cox
Director of Concept Vehicle Design,
General Motors[i]

Automotive design has traditionally been catholic – drawing influences from a wide variety of sources. While the self-referential nature of the motor industry has never been denied or concealed, drawing on the past has become an ever more explicit and acknowledged source of inspiration. Car design begets car design, and automotive designers are themselves the most devoted consumers and promoters of the iconic images of the past, along with the mystique that accompanies them. Asked what inspires them, designers conjure up elements of vehicles past and present, from details as small as door-handles and plastic seat wrappers, to the fabled design philosophy of celebrated marques. The hermetic world of the motor industry, from education through to career, coupled with the tangled web of allegiances, technology partnerships, and ownerships that characterize the modern industry, also makes self-reference inevitable. Just as fashion, architecture, music, literature, and design have all learnt to quote liberally from the past, so car design has generated its own visual language, a semiotic wonderland of signs and signals that defines each and every brand with an image and an expectation. To master this language is the car designer's ultimate goal, and every manufacturer strives to present a range of cars that demonstrates a cohesive expression of this elusive automotive alphabet, a cultural by-product of the automobile century.

Brand Identity

Perhaps more than any other industry, today's car companies realize the need to maintain and develop their brand identity. The brand – that mythical, intangible device that has become the primary item of currency in a global market built on the consumption of image and identity – is best expressed in automotive terms as a company's design DNA. The industrial designer Richard Seymour, whose consultancy Seymour Powell does considerable work for the motor industry, has written about car designers' quest for "authentic DNA,"[ii] decoding and combining elements of past production to create the most flattering public image of a particular marque. These DNA strands are often widely dispersed, ranging from the semantic to the subconscious. A famous name (Thunderbird, Speedster, TT) might be revived or placed in a new context. The shape of a vehicle – long, low and lean, or sculpted, curved and bulbous – has the obvious ability to make associations with past models, even past eras, as well as the harder to analyze political and personal symbolism. Or the strands might be found on a macro level in a small item of detail – a tail fin, porthole, chrome strip, trim, rocker switch, or name embossed on a kickplate: all elements that can be revived and re-presented. Similarly, there are negative elements which must be avoided – infamous cars, disastrous details, failed technologies, etc. – and purged from the present. Put simply, correctly decoding one's design DNA might create a vehicle that is a modern interpretation of

a classic vehicle from the past, appealing to a market segment that, while not explicitly wedded to nostalgia, is looking for today's technology combined with classic style. A company such as Alfa Romeo uses this concept – for example, the 1996 Nuvola – to decode key elements of the company's sporting heritage. The Nuvola shaped the company's current design direction, re-emphasizing the heart-shaped Alfa grille, lost in the angular frontal treatment of the 1980s.

The search for these strands of design DNA finds its purest expression in the concept car. Renault Senior Vice President Patrick le Quément's assertion that the concept is *haute-couture* while the production car is *prêt-a-porter* holds true. As in fashion, working on the cutting edge is the ultimate goal of every concept designer, pushing demands on materials, forms, and technology. Typically, car interior and exterior design departments are separate, each reporting to the engineering department for details on how changes and features will alter production times, material supply and, of course, cost. Margins are tight: allowing more variety in the consumer's choice of interior trim might have grave knock-on effects for the just-in-time production systems in use at the factory: if a component isn't supplied to the right part of the line at exactly the right time, it won't get fitted and delays will occur. The concept car is largely free from these restrictions, a complete work designed by one small team.

The bottom line is still money. Concepts – expensive, hand-wrought objects – aren't just cobbled together for the pleasure of a select few designers. Their vital marketing and research role demands that the business of concept design is serious and structured. The majority of concepts displayed by the major manufacturers in the past decade fall into just two categories: the development or revival of an existing brand, or a pioneering incursion into an all-new market segment. The former has risen to prominence in recent years.

The Past Reinterpreted
Retro design, as many commentators have christened it, isn't simply about plundering the past. Few car designers will admit to creating a "retro" design as such; the past is for inspiration, not slavish imitation. Modernist critics argue that "retro" patronizes the consumer, presenting design as a form of entertainment, devoid of progress. Admittedly, our preconceptions as to how a new car will perform are now largely driven by design. It's said that car companies can't afford to make a bad car, only a boring one. It therefore makes sense for car companies to stress difference and individuality through styling. However, even if a new car looks old, we have an inherent belief that it will be as safe, entertaining, and efficient as anything on the market, if not more so. It was obvious that J Mays' 1994 Concept One for Volkswagen, which eventually became the New Beetle, wouldn't rattle like the original bug, or limp up hills,

1994 VW Concept One:

J Mays' original New Beetle concept reduced the original to a series of geometric shapes, topped off with quirky touches like the on-dash flower vase (carried over to the 1998 production model).

handle erratically and have a rasping, raw engine note. Styling therefore becomes more than added value. The visual allusion to the familiar that cloaks contemporary technology speaks volumes about who we are and how we see ourselves. In short, we search for cars that are expressions of our personalities, and the dream of the concept opens up whole new arenas of individualism. Faced with an educated and demanding consumer base, car companies hone and direct our instincts, using concept cars as the lure.

American manufacturers lead the way in re-packaging the past as a gift to the present. As well as an Advanced Design Studio in England, headed up by RCA graduate Simon Cox, General Motors has a dedicated Corporate Brand Character Center, devoted to identifying how each individual brand is conveyed, and which would benefit from modern interpretations of the past. Most manufacturers have design centers located in each of their primary markets, sounding out new directions. Toyota's European Office of Creation (EPOC) opened in Brussels in 1989, while almost all European and Japanese manufacturers have footholds in North America. Global studios act almost like student campuses, centers of creativity that soak up the local culture and feed it back to their "square" parent companies. BMW's Director of Design Chris Bangle decided to use their semi-independent American studio, Designworks/USA, helmed by Adrian van Hooydonk, as the generator of BMW's current design direction.

This was spearheaded by the X-Coupé, and seen in production on the latest 7-Series, and is partly about decoding and integrating suitable American design values into a German company.

There is strong public enthusiasm for heritage mining. In 2003, Chevrolet will launch the Super Sports Roadster (SSR), a pick-up infused with 1950s bulbosity. The SSR was derived from the 2000 concept vehicle, quoting a style which, like the 2000 Buick Blackhawk and the 1995 Chrysler Atlantic, dates back to before the era of fins. "The Chevy SSR evokes Chevy's rich heritage," begins the promotional literature, explaining how the concept draws inspiration from the past, with pronounced wheel arches, chromed front and curves[iv]. Positive public reaction gave production the green light. Not all designers concur that this is retro design, preferring to see revivalism in terms of bolstering brand identity.

"For too long, the auto industry has been hermetic, looking to itself for inspiration," noted J Mays, Vice-President of Design at Ford, in an article setting out Ford's design ambitions (briefly summarized as "consistency, clarity, intelligence, style")[iv]. Mays claims to be looking to broaden the company's horizons, although he also claims that, "There is no shortage of heritage cues to exploit."[vii] More importantly, Mays states that, "Car design is converging with product design... the Ford Ka... was the first car to jump the species barrier

AIRBAG

and be more like a product than an automobile." Many would argue that cars have always been products, but the revealing insight of this line of thinking is that it represents a fundamental shift of emphasis within the industry. Mays talks explicitly about aping the effect "brilliantly designed" products have on consumers, products that project an image of tactility and desirability. He believes this must be done through creating a consistent visual identity, similar to the approach of manufacturers like BMW and Audi, with only subtle improvements that "flatter the consumer" through acknowledging the superiority of their taste. Mays, who worked at VW/Audi from 1984 to 1995, is a master of this "retro-futurism." With a dedicated exhibition at the Los Angeles Museum of Contemporary Art under his belt, he has overseen a series of striking, yet reactionary models, including Ford's Forty-Nine, with overtones of the 1949 model, and the new Thunderbird (manufactured from 2002 after the 1999 concept).

Globalization

The motor industry is at the forefront of globalization, and design centers are international businesses, spread out across continents in order to tap into the demands and quirks of the many different markets and their requirements. Nissan's US design studio, Nissan Design International, was established in 1980 in La Jolla, California. The West Coast is a popular location for international studios – more than 20 companies have facilities there, including DaimlerChrysler,

Mitsubishi, Honda, Volvo, and Mazda. A design studio is also, to a certain extent, a status symbol and is frequently located in so-called cultural capitals – London, California, Barcelona – all the better to soak up the creative atmosphere. "They feel the need to be in a cultural center, not a manufacturing center,"[vi] according to the RCA's Dale Harrow, who believes there is a cultural shift underway, as specific personalities become associated with certain designs, reflecting the consumer desire for individuality. With a relatively small number of big players, each controlling a huge number of often disparate, even competing, brands, management of design is crucial. Industry watchers talk of market segments, trend lines, crossover vehicles – an indication of the huge emphasis placed on design.

Personality Motoring

In this brave new world of brands, products, and emotional connection with the consumer, the most significant Ford concept – certainly the one which made the best connection to the world of product design – was the 021C, designed by the Australian product designer Marc Newson. Widely touted as the shape of things to come by the style press (who admittedly didn't know any better), the 021C debuted in vivid orange and lime green, at once distinct from the sober silvers, blacks and grays of the high-end concept, and therefore modern, while tipping a knowing wink at the playful colors and shapes of the plastics era. Inside, luggage and seating

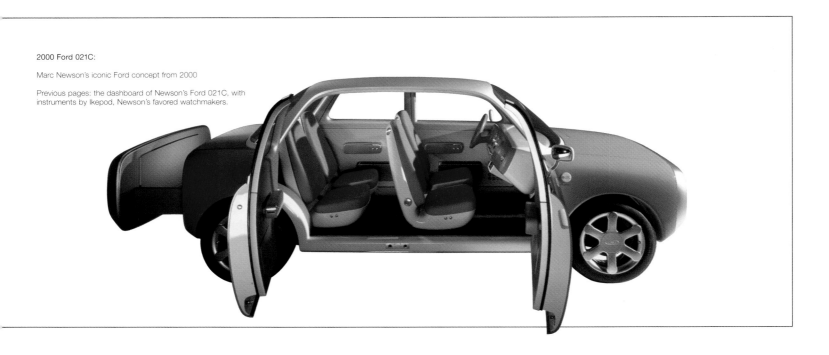

2000 Ford 021C:

Marc Newson's iconic Ford concept from 2000

Previous pages: the dashboard of Newson's Ford 021C, with instruments by Ikepod, Newson's favored watchmakers.

are by B+B Italia and the instruments are by Newson's own Ikepod watch company. In silhouette, the 021C is perhaps closest to the 1966 AMC Cavalier, an innovative early attempt to create a vehicle with interchangeable body panels, thanks to the symmetry between the front and rear of the car. For all its references to the design world, the 021C will remain a concept – the technology is just not there to realize it (neither are the regulations that would make it legal). To the visually literate, the 021C looks a little old-fashioned and strangely familiar, just like a sci-fi vision of the 21st century (not surprising coming from a designer who once chose a space suit as one of his enduring design icons[vii]).

Freeing car design, and concept design in particular, from our imbedded cultural expectations is all but impossible. For this reason, the sources of inspiration tend to be familiar and tangible, aspirational yet not exclusive. Jaguar's R-Coupé and the Lincoln Mk9 (produced by design teams headed by Ian Callum and Gerry McGovern respectively, and both coming under Ford's global umbrella) illustrate how the classic status indicators of performance and size can be melded with contemporary design acumen to create futuristic vehicles which draw strongly on the past. Both are studies of future "design direction," big coupés for big spenders, concepts to inflame desire and boost the standing of their respective brands.

Nissan and the 'Z' Series

Nissan recently revived their "Z" series brand with the Z-concept, due to reach production in 2003. A presentation given by Thomas H Semple, President of Nissan Design in America, reveals the thinking behind the concept[viii]. Asking, "How can we reach back to our heritage?," Semple proposed revisiting the "Z"-series, the brace of sporting coupés built by Nissan from 1970 onwards. Recognizing the growth of the coupé market was one thing, but the new vehicle involved in-depth consumer research, "to see if there was still a dream." The sketching and 1/4-scale model-making process saw a number of minimalist solutions proposed, and by the time the new "Z" had reached the 1/1 model stage, three months later, a clean coupé shape had emerged. The Z's flanks were carefully creased, with wheel arches flaring out and intersecting with a muscular curve that bisected the door. The whole ethos of the "Z"-series – a sporting coupé initially intended to go up against the floundering American muscle cars in the early 1970s – is ably summed up by its silhouette, a shallow curve that extends from bonnet to boot and is then inverted and carried through to the interior, forming the sloping central instrument console. Once the interior and exterior had been selected – in March 2000, ten months of production started, leading to an initial show car at Detroit in January 2001 and a more developed idea later in the year, at Tokyo.

2000 Cadillac Imaj:

An early "art and science" design, the Imaj was far more progressive. Billed as a luxury sedan, the Imaj was a technological showcase, stuffed with advanced communications and entertainment systems.

2001 BMW X-Coupé:

This all-wheel drive concept formed a distinct departure from BMW's in-house style. Calling the bodywork treatment "flame surfacing," it is a composition of arcs, bulges, and intersections. The vast, lop-sided tailgate provides access to the trunk.

Expression

2001 Jaguar R-Coupé:

Ian Callum headed up the design team that
produced this classically simple coupé concept.

1992 Vovo ECC:

This model introduced a new image for Volvo, exchanging boxy angles for a more curvy sihouette.

The Z-concept is a good example of how inspiration develops and changes during a concept's genesis. From the decision to revive the brand, through focus groups and intensive design, the finished vehicle was not just a simple throwback. Key design elements, such as the scooped instruments – a detail which is carried over into the door grip – might have initially derived from the original 240Z's headlight arrangement, but have been allowed a new lease of life elsewhere on the model. Similarly, the three generations of concept, developed over an 18-month period, saw the design alter and accommodate new features, all part of the process of sounding out potential production worries. As the next chapter will demonstrate, bringing a concept to market is a complex process, one which needs to be considered from a very early stage.

Freedom of Expression
Away from the world of motor shows and million dollar budgets, design inspiration is unhindered by rules. Student work can, and frequently does, use transportation design as an improbably broad canvas, blending genres and styles, concepts that would trouble even the most forward-thinking marketing department. Briefs allow for the use of yet-to-be-developed materials and drivetrains, developing vehicles that are amphibious, biomorphic, environmentally aware, for roads that are a quarter-century away. Sponsorship, which plays an important role in developing student work, means that expression

is frequently channelled through industry-defined routes. Towards the end of 2001, Detroit's College for Creative Studies ran a design study sponsored by PPG Industries, the largest automotive glass supplier in North America. Students were asked to project future uses of glass in a vehicle, imagining how a high-end 2008 model might use the properties of glass to enhance safety, security, and comfort.

Brand Definition
Concepts can be used to reinforce a particular trait, one which might be losing definition. A more grounded example of the potential safety role of glass is demonstrated in Volvo's 2001 SSC Safety Concept Car, styled at ItalDesign's California studio. Finished in high-visibility day-glo orange (less a fashion statement, one suspects, than Newson's Ford), the SCC was described as "first and foremost, a safety concept to re-establish Volvo's safety credentials."[ix] In other words, the concept allowed the company to be more aesthetically adventurous, addressing the staid image of its saloons without compromising its reputation. Volvo's status as the most safety-conscious car-manufacturer saw a similar exercise in the 1970s, when an earlier safety car concept was manufactured as the 240, the sharp-edged model that introduced the famous boxy style (since usurped by the curvy aesthetic first defined in the 1992 Environmental Concept Car [ECC]). The SCC's innovations include a four-point safety-belt and semi-transparent A-pillars.

1995 Chrysler Atlantic:

The large, two-seater Atlantic quotes several cars, most obviously the Bugatti Type 57S Atlantic, the Talbot lago Coupé and the classic postwar Jaguar XK120. The company described it as "classically-inspired."

Trends of Expression

Like fashion, sources of inspiration move in cycles. The concepts of the late 1960s and early 1970s looked laughably angular when compared to the more rounded sensibility that came to the fore in the 1980s. Yet today those angular shapes are starting to prove inspirational themselves, reproduced in glowing spreads for fashion magazines, catching a second wave of admiration. As environmental concerns grew in the public's consciousness, angularity gave way to insectoid, organic design, coupled with exploration into forms compatible with new, greener power sources. Peugeot's Tulip city car – the work of maverick Swiss designer Luigi Colani – the electric GM Aero, and the Buick Questor concepts of the early 1980s, all shunned the straight line for a friendlier curve. IAD's maverick Alien concept (1986) epitomized this trend, a bulbous passenger cabin attached to a power plant. Fiat's 1996 Armadillo took organic design to extremes, expanding the car's inherent anthropomorphic qualities into a sleek, cute bodyshell. These were just the models that made it to concept stage – never mind the countless student projects that imitated animals, insects, fish – a veritable automotive menagerie.

Laboratories of Style

Today, the cynic might be forgiven for thinking that artistic inspiration is sold short by many of the concepts produced by the major manufacturers. The swift turnover and high volumes make it easy to forget individual examples, turning the concept car into little more than a passing entertainment, a popcorn double feature. Obviously concepts aren't merely for show, regardless of the level of fanfare and attention they receive on the podium. Studios might produce concepts that never see the light of day – secret styling exercises for board presentations and in-house design clinics. The industry's far-sightedness is often underplayed; design studios are typically working three to five years ahead of today's market, and the concept designer is at least five to ten years ahead of them. For all this, the development period for a concept is usually far less than that of a production car – perhaps only a couple of months as opposed to a minimum of two years. In this respect, the concept is a sketch, albeit a highly-worked one, allowing it to be an expression of anything from new technology, new style, or just a means of creating an emotional bond with potential consumers.

The concept car was initially devised to creep into the corners of the consumer psyche, causing fresh associations and stimulating new demand; hand-built to manufacture mass-market desire. Regardless of the level of technological innovation involved, concepts are

arguably better marketing devices than technological testbeds, functioning as laboratories of style. This is in a market where mechanical similarity is engendered by a globalized, multi-owner car industry, and styling is elevated to hitherto unknown levels of prominence. The car market is built on brands, making the upper layers of an automobile – of metal, plastic and chrome – the place for defining presence. And increasingly, concept cars are becoming real products, as the increased flexibility of production technology and the pinpoint accuracy of market research helps them make the move from the motor show to the showroom.

Footnotes

i   Simon Cox quote on p148; 'Reviving a Great Tradition' by Kate Trant, Moving Objects: 30 Years of Vehicle Design at the Royal College of Art (RCA Publications, 2000).

ii   'Car Genetics', Domus magazine, Richard Seymour (September 2000).

iii   Likewise the Dodge M80, with its Jeep-like radiator treatment, prominent wheel arches (painted a different colour from the bodywork), and tight, tucked-in passenger cab, emphasizing separation from the chassis and flatbed, also comes from the same era.

iv   p44–45, 'The Vision Thing', Blueprint magazine, J Mays (February 2002)

v   Ford's recent heritage inspired concepts include the Ford Forty-Nine, the new Thunderbird, a revised GT-40, and the Lincoln Continental and Mk9. Ford's iconic cars include the 1955 Thunderbird, 1964 Mustang, the original GT-40, and the 1961 Lincoln Continental, a black slab of automotive modernism overseen by George Walker, perhaps the company's most influential design consultant. Ford's Thunderbird (now in series production), Forty-Nine, and GT40 are all symptomatic of this attitude. Under the design leadership of J Mays, formerly VW and Audi (he described the latter's approach to design as ''elitist''), this trio of concept vehicles (the Thunderbird has now entered series production) all owe deep debts to the 1950s – the industry's self-declared golden age.

vi   In conversation with the author (February 2002).

vii   Marc Newson's Conran Collection, Design Museum (2001).

viii   Presentation given to the AutoWeek Design Forum, Detroit (2001).

ix   Design Director Peter Horbury quote on p39, 'Press to Play' by Amelia George, Volvo Magazine (Spring 2002). More detail can be found at www.conceptlabvolvo.com.

The Fiat Armadillo, 1996:

Designed by the company's in-house Centro Stile in conjunction with specialist coach-builders Maggiora, the Armadillo is a low-drag, anthropomorphically-styled city car with a flexible, spacious interior and a friendly appearance.

Classic styling
Citroën

Citroën C-Airdream:

The 2002 Citroën C-Airdream, the company's dramatic concept coupé, melds new technology (drive-by-wire and other electronic control systems, including steering-wheel mounting brakes) with a shape that develops Citroën's classic teardrop shape for the 21st century.

Classic styling
Citroën

Traditionally, concept cars represent a design extreme, far removed from pragmatic reality. Yet all that is changing as manufacturers take on board the admiring glances that consumers give to concept cars as the demand for design individuality increases. Today, it is difficult to find a genuine concept amongst the shiny new models at a motor show. Are concept cars examples of pure speculation or simply a preview of next year's model?

One company aware of the need to project a stylish future is Citroën. Once upon a time, the French manufacturer thrilled car buyers with ayseries of audacious designs that were as technically innovative as they were physically striking. The original DS, unveiled to an awestruck public in Paris in 1955, and its subsequent replacements, the CX and XM (not to mention the SM and GS models) remain unrivalled for their unconventional style and reliance on hydraulic systems for steering, braking, and suspension. After being swallowed up by rivals Peugeot in 1974 to create PSA Peugeot Citroën, Citroën began a slow decline, and the company squandered its reputation for innovation through successively blander production models.

2002 Citroën C-Airdream:

This coupé showcases drive-by-wire control system.

Citroën's recent concepts have been an attempt to redress the balance. 1988's Activa concept, for example, revisited the combination of unconventional technology and eccentric styling, a pairing successfully continued in the 1999 C6 Lignage, an in-house design by the company's Creative Styling Center. The Lignage's streamlined flanks didn't materialize in production form, although one can perhaps see a debased version of its silhouette in the current Xsara Picasso. However, in the past two years, Citroën have embraced the concept car as production prototype, demonstrating their willingness to innovate with the transformable C3 Pluriel, a small car that offers multiple configurations through removable body panels that re-shape the load and passenger areas: from hatchback to pick-up to convertible. Shown first as a concept at Frankfurt 1999, a production version debuted at Paris 2002. The detachable roof components are fabricated from aluminum, making them light and easy to store.

Also shown at the Paris show was the C-Airdream, a sleek 2+2 coupé that harks back to Citroën's earlier, teardrop aesthetic, along with the Hydractive suspension that has remained in continuous development at the company. A glazed roof gives the car an airy interior, and the C-Airdream is yet another concept that showcases drive-by-wire controls, an innovation recently trailed by both Bertone and General Motors. (Interestingly, the original 1950s DS replaced the brake pedal with a rubber button, so efficient was the hydraulic brake system). The forward thrust of the lines – which emphasize the long bonnet and tapering rear – are classic Citroën styling cues and bode well for a future return to the more individualistic end of the market.

Expression

Designer as maverick
Luigi Colani

Luigi Colani with his make-over of the Ford Ka:

The original Ka, introduced in 1996 after the concept was shown in 1994, was styled by a team led by Chris Clements. Colani's bodykit shows the application of his baroque organic aesthetic to a mass-market vehicle (one which was already a highly competent visual package).

Designer as maverick
Luigi Colani

Luigi Colani has a reputation as a maverick, an industrial designer operating on the fringes of the profession. As a result, this Swiss-born designer is something of a cult figure. After travelling widely, including time in Japan and France, he eventually chose to settle in the German town of Fuchtorf, combining home and studio in a local castle.

Colani's vehicle design exists in another sphere from the marketing-driven concepts of the major motor manufacturers. A lifelong interest in transport design has resulted in automotive projects like the Colambo and the Al-4 – sleek, fish-like bodywork encases a highly styled passenger compartment, with almost recumbent seating. Colani termed this approach "bio design," steadfastly refuting the dominance of the modernist aesthetic and spending his spare time researching organic forms for inspiration. His steroidal, mechanistic art nouveau is highly original, grounded in sound aerodynamic and ergonomic principles, creating concepts – before the term was officially coined – that appear fresh today. In 1979, asked by BMW to project a future look for their 5-Series saloon, Colani came up with a

Horch coupe 1996:

This streamlined coupé built in 1996 was Colani's attempt to
revive the 'Horch' brand name, once a leading player in the
German auto industry. The concept's long front and rear
overhangs as well as the sinuous line of the bodywork
are all the designer's hallmarks.

streamlined proposal years ahead of the market, even after two
decades of design evolution on the production model. Unperturbed
by the company's disregard for his suggestions, in 1982 he retorted
with the outlandish BMW-M2. A curvaceous companion for the
company's 1980 M1 model, it was an ItalDesign-created supercar.

Even though Colani's studio could produce the Citroën 2Ch, an
aerodynamic concept vehicle that was achieving 21st-century levels
of fuel consumption back in 1981, occasionally his mindset resulted
in wilful and eccentric anachronisms. In the 1970s and 1980s, Colani
projected a future of automotive styling that never came to pass,
extrapolating a weird, parallel aesthetic derived from the classic
pre-war styling – all faired-in wheel arches, independent body and
roadster styling. The 1976 Aiglon is the peak of this quest, a vast
two-seater that epitomizes this "neo-classical-style research."

The inside-out approach, starting with ergonomics and working
toward enclosing the passenger space as efficiently as possible,
means that Colani's concepts are fully functional – often using donor

production cars as an initial starting point. Despite consultancy work
for NASA, Colani became disillusioned with the conservatism and
lack of innovation in Germany in the 1980s, especially the motor
industry (he once called the automotive business "one of the most
conservative and reactionary industries in the world"). He made for
Japan, where he became hugely influential on consumer product
design. He is now back in Europe and still working. The mainstream
creeps ever closer to Colani's aesthetic.

The advanced concept studio
Renault

The 1992 Renault Racoon was a return to the traditional "fantasy" concept, an amphibious off-roader with infra-red night-driving abilities and go-anywhere capability.

The advanced concept studio
Renault

Renault, under the design leadership of Patrick le Quément, has created a varied set of concept cars in the past two decades, straddling the line between shrewd market-research and outré fantasy. Unusually for a major mass-market manufacturer, however, Renault concepts have had more than a mere psychological effect on production cars. Having joined Renault in 1987, le Quément has arguably revitalized the company's design emphasis with concepts like 1990's Laguna Roadster. He has also led a revolution in production design, most notably with 1993's cheeky Twingo and the 1996 Sports Spyder, a windscreen-less roadster.

The link between concepts and Renault's production cars was initially tenuous. Proposals like the Racoon (1993) were far removed from production reality. An off-road concept produced first in virtual form to showcase the company's new modeling software, the Racoon doubled up as a watercraft, with passengers housed in a huge, aircraft-like dome. 1996's Fiftie was a deliberate attempt to recreate the past, a fresh interpretation for the 50th anniversary of the iconic 4CV model (in today's market, the unashamedly retro styling would

1994 Renault Argos concept car:

The Argos, first shown at the Geneva Show, was a pared-down evocation of the Renault brand's
key points: simplicity, sportiness and a commitment to strong modern design – an "esprit nouveau."

probably be a roaring success). These were cars infused with a spirit
of fun, inspired by the shiny utopian surfaces of science fiction.

As the 1990s progressed, Renault became more explicit about the
role of concept cars as signposts for new design directions and new
market segments, expressing a new luxury aesthetic through
unconventional angles and body shapes. Most radically, 1999's
Avantime concept was proposed as a luxury coupé MPV, a hitherto
improbable mixture. To the surprise of the industry watchers, Renault
announced that they would build the Avantime from 2002, in a form
which differs only very slightly from the original concept.

The production Vel Satis was also dramatic, although relatively sober
compared to the concepts that inspired it, notably 1995's Initiale and
the original 1998 Vel Satis concept, with its neatly recessed
glasshouse and angular front and rear treatment. Vel Satis is also
about market positioning – it is Renault's stake in the luxury market.
While the company has always built big saloons – the 25 and the
Safrane – it has been unable to compete in prestige terms with high-
end German and British marques. In the absence of an existing
customer base, Renault's trip upmarket therefore began on the
drawing board, beginning with 1988's Mégane (the company
recycles concept names for production cars, regardless of any
physical connection – there has also been a concept Laguna (1990)
and Scenic (1991), although the latter was a mini MPV). Concepts
expressed a feeling of quality that was lacking in production models.

The success of the company's decision to bring the avant-garde
to market is still an unknown quantity. Both the Avantime and the
Vel Satis represent radical departures from current production
models, thanks to a policy of using concept cars to showcase not
only new design directions, but also open up new market segments.
The 2001 Talisman concept went further – a luxury four-seater GT,
with two vast, gull-wing doors and a luxury interior designed to lure
the BMW driver.

The Lincoln MK9:

Design sketch for the 2001 Lincoln MK9 concept, a long,
low coupé that harked back to the "high" era of late
1950s and early 1960s American car design.

American luxury:
Lincoln

The 1996 Lincoln Sentinel:

First shown in 1996, the Lincoln Sentinel concept was the start of this seminal American brand's attempt to move upmarket. Sleek, crisply detailed, and ever so slightly menacing, the Sentinel strove to bring the classic lines of the 1960s Continental model back to a line-up that had grown staid and lifeless.

Few manufacturers have been as explicit about the link between past glories and new concepts as Lincoln, Ford's luxury American subsidiary. The company that created the celebrated Futura in 1955 (styled by Ghia and later to find global fame as TV's "Batmobile"), Lincoln's recent mission to reinvent the past has taken modernism and heritage as its major cues. Their first foray into retro concepts was the 1996 Lincoln Sentinel, presented with the slogan "what a luxury car should be" and exhibited at the UK's RCA "Moving Objects" show the following year. If anything, the Sentinel looked and sounded a little sinister: subsequent concepts have retained this severity, but have also emphasized a link between heritage and innovation. The Sentinel was followed by the MK9, first shown at the 2001 New York motor show and hailed by Lincoln as the ultimate re-interpretation of Detroit's past glories. The MK9's "American Luxury" made deliberate reference to contemporary design and architecture, positioning America at the forefront of the modern movement's transition to international modernism. The slabbed design of the car's body, reminiscent of the architecture of Mies van de Rohe, locates the Lincoln alongside the elegant, barge-like creations of the 1950s and 1960s, modern classics that were also loaded with innovation and, more importantly, luxury.

This direction was developed further with a new Lincoln Continental concept at the 2002 Detroit Show. Although the first Continental was introduced in 1940 as a European-inspired model, the car swiftly came to epitomize the upper end of the American market, outgrowing its fins and evolving into something long, low, and sleek. The new Continental concept is fashionably squared off, making the 2002 production model look almost baroque by comparison. The sleek bodyshell opens up, clam-like, with an extending trunk reminiscent of Newson's Ford 021C. Under the design direction of Gerry McGovern, Lincoln has used concepts as brand reinvention, producing designs that raise the company's profile (and with it the image of current production models), and reminding customers old and new of a carefully controlled corporate heritage.

American retro:
Buick

Buick's Blackhawk, introduced in 2000, is the ultimate retro machine, drawing styling elements from the company's 1939, 1941, and 1948 cars, to create a vehicle steeped in the imagery of pre-war America. Other recent Buicks, such as the LaCrosse and Bengal from 2001, have quoted freely from the past (in particular with the LaCrosse's Y-Job-inspired frontal treatment), but none are as blatant as the Blackhawk. Buick's parent company, GM, has an enormous amount of visual material to draw on, as well as a hefty portfolio of distinct brands: Buick, Chevrolet, Cadillac, GMC, Oldsmobile, Saturn, Pontiac, and Hummer in the US; the EV1 electric car program; the related Vauxhall, Holden, and Opel family; and Saab. Each has their own markets, identities, and consumer expectations, a brand spread that allows for the promotion of either pure futurology or unashamed heritage design.

Buick:

The 2000 Buick Blackhawk concept. With a grille strongly influenced by the 1939 model and lines derived from the company's 1940s models, the Blackhawk is thoroughly, unashamedly retro.

Japanese style, international outlook

Isuzu

Isuzu:

The 2000 Isuzu Zen fused elements of traditional Japanese
domestic architecture with contemporary automotive
technology to produce a surprisingly forward-looking concept.

The Isuzu Zen concept debuted at the Tokyo 2001 motor show.
Intended to appeal to a new market segment, with a mix of SUV and
the ubiquitous panel van features, the Zen is a study in automotive
serenity, a Shinto temple on wheels. While the rugged exterior is
reminiscent of metal-clad industrial buildings, it is the Zen's interior
that make these Japanese influences explicit; a spacious, fold-flat
passenger compartment that embodied the design team's slightly
risqué maxim, "anything, anywhere, anytime and anyone." Sliding
side windows are allegedly inspired by the traditional Japanese
folding fan, or "ogi," interior surfaces are finished with tatami mats
while the tailgate, which raises vertically into the bodywork, is
directly lifted from the traditional translucent *shoji* doors of Japanese
architecture (especially the "yukimi" *shoji*, or "snow-viewing" doors).
Unusually, given its inspirations, the Zen was designed entirely in
England, under designers Geoff Gardiner and Ian Nisbett, the same
team responsible for the more aggressively styled Kai (2000).

"No matter what technology is available, unless a car is carrying a load of effective imagery it is unlikely to enjoy genuine popular success."

Stephen Bayley
Author of Sex, Drink and Fast Cars, 1986

The marketplace is choked with cars – nearly 300 individual models. Each manufacturer peddles a fleet of different models, each offering a dizzying plethora of options, accessories, finishes, and special features. No two cars need be alike. This new era of choice and availability, not to mention rising sales – exceeding 200,000 a month in Britain for the first time in 2002 – has made differentiating between individual models the industry's primary goal. We need to feel that the cars we buy are suitable expressions of our values, aspirations, and taste. Selling cars is big business – in 2000, General Motors had a $2.8bn advertising budget, the largest in corporate America. Motor industry advertising and sponsorship pervades every pore of our society. Car culture is deeply ingrained. Image is everything. "No matter what technology is available, unless a car is carrying a load of effective imagery it is unlikely to enjoy genuine popular success," notes Stephen Bayley[i].

Communicating Core Values

Concept cars are an extension of that imagery; virtually real, highly evocative vehicles that stimulate our desire to buy and drive. Bob Lutz, GM's vice-chairman of product development from August 2001, once said that, "The ability to inspire passion is the single most significant measure of a vehicle's success."[ii] Lutz went on to explain that design is the pre-eminent factor in communicating a vehicle's "index of passion... real differences in quality are so narrow as to be almost

negligible." Defining design is complex. It is a word that can be shaped to almost any end, molded to fit any ideology or idea. Bayley tells of how Audi's British advertising agency, BBH, pulled off a masterstroke in the 1980s by reviving the slogan "Vorsprung Durch Technik" for the UK market, transforming a dour but efficient saloon car manufacturer into a widely perceived powerhouse of modernist elegancy and functional integrity. "Vorsprung Durch Technik" became a self-fulfilling prophecy, and design became the Audi brand's big selling point.

It is transparently obvious that cars are highly-designed objects. However, the public's visual awareness levels have risen steadily, as has the consumer obsession of the age: branding. We are obsessed with brands; their meanings, values, and identities. Suddenly, the role of consumption in creating a sense of social identity has never been so blatant (although design theorists have been highlighting the link between consumption and identity for decades). The brand is the pre-eminent marketing concept of the 21st century, and it seems only natural that automobiles should have evolved as the ultimate expressions of brand identity. Nothing has proved more effective at communicating a car manufacturer's core values than the concept car. While at Chrysler, Lutz marketed the highly popular retro-styled Dodge Viper, Plymouth Prowler, and PT Cruiser, all radical concepts that made it to market. Believing strongly in the existence of a right

automotive aesthetic – "the body should look like it was stretched
tautly over the chassis" – Lutz's design-focused approach has been
immensely influential (Lutz himself has described chief designers
as overseeing a "benevolent dictatorship"). The public's ability
or inability to recognize or endorse a brand can make or break
a car, so extending a manufacturer's range has to be handled with
care. Concepts play the part of advocates, enforcing, even creating,
brand values as well as pointing to the future.

The Designer as Hero
Celebrated early concepts, such as Harley Earl's wild Buick Y-Job,
built in 1938 were very much stamped with an individual identity,
regardless of the size of the team used to create them. The work
of studios like Pininfarina, Zagato, or Bertone also bears the stamp
of individuality, with certain motifs, devices, and shapes repeated
until they become familiar – Pininfarina's razor-sharp creases, or
Zagato's muscular bodywork, with the twin-bubble roofline, for
example. These signature forms made the products of these studios
immensely desirable, establishing their proprietors as the masters
of car design, lionized by enthusiasts and the motoring press.
Car styling also has its new stars, as those who understand brand
values and how to distil and develop their key characteristics rise
up to positions of power. Designers like Gerry McGovern, J Mays,
Freeman Thomas, and Walter de Silva are people for whom style

magazine spreads and messianic proclamations are all part of the
promotional push. Nissan and Audi designers have even appeared
in television commercials, and are wheeled out to talk to the press
about their passions and inspirations[iii]. Not since the heyday of
Raymond Loewy, Normal Bel Geddes, and Harley Earl have
individual designers been afforded such prominence.

The Lure of the Future
Concepts are presented as the industry's *haute couture*, expensive
one-offs that strut the motor show catwalks to showcase a brand. This
fashion industry analogy is still pertinent. Concepts once brought an
almost politically incorrect element of fantasy to the motor industry,
far removed from the average family car, a hedonistic, forbidden
symbol of recreation, entertainment, and pleasure. In the 1950s,
concept cars allowed us to buy into the future. They led us toward
tomorrow. On the face of it, today's concepts bear little, if any, relation
to the concept cars – or dream cars – of the post-war era. Carefully
stitched together from the genetic material of their forebears, the
concept car is backed up with technology that ensures it can make
a near-immediate impact on mass-market designs. Contemporary
motor shows are littered with new models, all promising hitherto
unavailable lifestyle combinations – sportscar/SUV, city car/off
roader, luxury/estate. For the consumer, concepts are not simply
extensions of our present: they are things we expect to become real.

From Concept to Production

Technological advances allow the concept to move swiftly from podium to showroom as supply, costs, and manufacturing issues are integrated into the early stages of the development process. Time is of the essence. Marketing a new car is dependent on speed and buzz. Once a project has been green-lighted, the bare minimum time to tool up and manufacture a mass-market car is around 12 to 18 months, a radical improvement on the four to five years of just two decades ago. GM, amongst others, use an all-digital process to the marketing stage, a system which crucially allows for the synchronization of quality control, manufacturing, and testing issues before any metal is bent in anger. Co-ordinating this kind of operation required an intensive overhaul of the company's corporate culture. A computer industry publication[iv] noted how the giant company turned itself around, replacing 3,500 IT systems that acted independently of each other, hindering efficient collaboration, and 23 separate CAD systems. With GM design centers located on 14 sites around the world and employing 18,000 designers, systems integration has been the single most important change in the way cars are designed and sold.

The nature of the contemporary car industry, with a massive pool of shared mechanical components (platforms) also hugely assists the speed of the process. As well as Buick, Cadillac, Chevrolet, Pontiac,

and Saturn in the US market, GM also owns Opel, Vauxhall, and Holden in Europe and Australia, Saab in Sweden and a 49 percent stake in Isuzu motors, a slightly smaller stake in Daewoo (Korea), and one-fifth of Fiat Auto. Cars that appeal to totally different markets and demographics – for example, the Saab 9-3 and the Vauxhall Vectra – might therefore share substantial mechanical components. Ford commands a global portfolio that includes Aston Martin, Land Rover, Lincoln, Volvo, and Jaguar, divided into mass-market cars and the Premier Automotive Group for the more expensive brands.

The upshot of this cross-pollination is that today's concept car is increasingly likely to be the production car of tomorrow. Pure concepts – those that act simply as beacons for the brand, unhindered by production considerations, like the traditional dream cars – are becoming few and far between. Audi's Rosemeyer Concept (2000), designed by Peter Schreyer and the Audi Design Team nods implicitly to Audi's past, especially the record-breaking Auto Union cars of the 1930s. Named after racing driver Bernd Rosemeyer, famed for his runs on the Avus raceway (and later killed in a high-speed crash on Berlin's Autobahn), the concept formed part of Audi's pavilion at Autostadt, the Volkswagen group's sprawling brand experience theme park at Wolfsburg. The Rosemeyer was exhibited in a mock-up of a motor show unveiling, all flashing lights and snatched glimpses following a short historical film: mystique and

AUDI Project Rosemeyer:

This is a dramatic supercar concept, a spiritual evocation of former racing glories. This and the AUDI Avus concept preceded the announcement of a production supercar, the RSR, due in 2005.

VW Microbus, 2001:

A modern interpretation of an acknowledged classic, the new Microbus draws upon the original vehicle's charm. Styled in VW's Simi Valley design studio, the Californian influences are obvious: variants of the concept featured a half-cab body and an accordion-type enlargeable roof.

myth blended together to enhance the brand's standing. Production was never considered.

For the most part, Audi's recent concepts have been closely allied to future production. Also in 2000, the company introduced the Steppenwolf, a sporting off-roader that combined Audi's famed four-wheel-drive know-how with clean, modernist styling. These elements were showcased in production vehicles such as the A6, an all-new incarnation of which was introduced in 2001, and the A8, overhauled in late 2002. The TT remains Audi's most famous concept. First shown at the 1995 Frankfurt Motor Show, the car was styled by Freeman Thomas (exterior) and Romulus Rost (interior). The TT won instant praise for its simple, uncluttered lines and influential interior (all finely honed dials, circular motifs, and lashings of aluminum). For Audi, the concept car was a means of expressing and reviving the core elements of its brand. The TT's visual dependence on strong, geometric forms created a convincing case for linking the firm with pre-war German modernism and the functional style of the Bauhaus in particular. The use of silver as a corporate color (since 1993) also draws parallels with the aluminum-bodied racers of the 1930s era. The past, present, and future are therefore brought together in a desirable package, elevating the consumer perception of the brand and, by osmosis, the other models in the Audi line.

Marketing Design Revivalism

The TT embodies the retro futurism discussed in the previous chapter. Although the TT adopted styling cues that had first surfaced in the 1991 Audi Avus, a J Mays-designed vehicle, the production TT was almost identical to the concept, the only compromises being a change to the intersection between the curved roof and waistline, necessary to facilitate production. One person's retro is another's brand awareness, and advocates of retro design argue that stylists and designers are simply addressing consumer demand for more distinctive and individual automobiles. Cultural memories die hard, and the familiar plays an important role: we are more likely to buy something which carries strong associations.

In June 2002 Volkswagen confirmed it would be building the Microbus, promising to remain faithful to the 2001 concept shown at Detroit. This concept had obvious antecedents: the subtle two-toned paintwork was no doubt informed by the lovingly restored examples from the 1950s and 1960s that congregate around the world's windier coastlines, or in flattened paddocks for sessions of mutual admiration. The new Microbus might be a million miles from its predecessor in all but spirit, yet it is precisely this spirit that has become the most important component to retain. This brave new world of design revivalism has its dangers. Funds are not always available to capitalize on the bold reinventions proposed by design

Lincoln Continental, 2002:

This concept was produced during Gerry McGovern's drive to re-brand Lincoln under the banner 'American Luxury.' Conventionally shaped, but with extraordinary attention to detail, quality and materials, the Continental was influenced by the model's 1960s and '70s heyday.

directors. Lincoln's loudly-hailed American Luxury strand, seen in the Sentinel, Mk9 and Continental concepts (see page 90) failed to make it beyond the motor show due to financial concerns.

Cautious Approaches

Today's car market is a seething confusion of strategic partnerships, technological collaborations, market segments, and names, marques and models. It is also a market steeped in self-awareness, the knowledge that the consumer now expects certain things from certain brands. A Jaguar must look like a Jaguar, a Volkswagen a Volkswagen, and design studies – concepts – that seek to show the very essence of "Jaguar-ness" or "Lincoln-ness" draw on classic cars from the 1950s and 1960s, the models deemed to represent the best expression of the brand. Manufacturers are acutely aware of the demands of their core demographic – extending their reach into new sectors is often a gamble. Jaguar might eye the young consumers snared by Audi's TT and Mercedes' SLK with envy, and recent Jaguar concepts appeared to encroach on the demand for small, sporting, yet upmarket coupés. However, market research has shown that it is vital for a Jaguar to be able to slot a set of golf clubs in the boot. Similarly, Ford concepts demonstrate a commitment to innovative, often avant-garde design, particularly through collaboration with its subsidiary, Ghia Design in Italy. Ford's status as a provider of value, mass-market design was once at odds with the radical styling

concepts demonstrated by Ghia (the 1993 Ghia Focus, for example, the single example of which sold for over $1m in June 2002).
It took the introduction of New Edge styling, a slow process trailed by several concepts before coming to full fruition in the Focus production car in the late 1990s. Now Ford claims to be a market leader in design: the recently opened Ingeni studio in London, a Richard Rogers-designed building in Soho, contains both car stylists and product designers[v]. So studios will be dedicated to producing branded projects – such as a lovingly crafted Aston Martin chess set, even a Volvo sled.

With concept cars functioning as the purest expression of a company's design intentions, their high profile now extends to their inclusion in advertising: witness Volvo's use of the Concept Safety Car in consumer publications. Alternatively, the difference between concept and production is deliberately confused. The Alfa Romeo Brera's promotional web site asks, "What difference does it make? What's the difference between a prototype and a series production car when you can move from one to another in the twinkling of an eye?" This conflation of concept and production is also way of pushing new technology onto the market. The era of atomic-powered cars has passed, and outrageously fantastical concepts make fewer and fewer appearances, much to the disappointment of enthusiasts. Automotive technology tends to improve incrementally, as

production, post-production and market problems are overcome. Audi's introduction of the aluminum space frame construction started with the A8 (having debuted as the ASF concept in 1993), right at the top of the market. To start with, only a limited number of high-end Audi dealerships had the tools and specialist knowledge to deal with repairing the cars' aluminum bodywork. With the introduction of the small A2 model (again, based on a concept, the Al2), the new technology went mass-market.

The Increasing Influence of Technology

Information technology and in-car electronics are also changing. Currently, regulations do not permit drive-by-wire steering systems: cars must have mechanical linkages between the driver and the wheels. Bertone, with their Filo and Novanta concepts, evidently believe that such regulations won't stay in place for ever. Each vehicle is a working showcase for drive-by-wire technology, with revised internal layouts exploiting the lack of mechanical linkages. BMW's controversial iDrive system, standard on the 2002 7-series and the 2003 5-series, was first revealed on the Z9 concept, a rolling test-bed of technology and style first shown in 1999. As the next chapter will demonstrate, concept cars that are based almost entirely around a technological concept rather than a stylistic idea are becoming increasingly common.

Marketing and Education

Students are made fully aware of the importance of the market and the role concepts play in reinforcing a brand. At the RCA, a 2001 course unit entitled Grafting Hybrids sought to discover exactly what brands mean. Having studied the meanings and associations of a series of high-profile non-automotive brands – such as Samsonite, MTV, and Apple – students were instructed to re-interpret these values in a car design. "Being able to capture some of this aspirational identity of a non-automotive brand and graft it into a vehicle concept, might be a low-risk way for car manufacturers to quickly and cost-effectively realize some of the potential of having a brand in tune with 21st-century consumers."[vi] The workshop produced some striking, but not improbable, hybrids. Eero Kankainen proposed an MTV/Opel taxi, while Andrew Sheffield took the tough, modernist image of Samsonite luggage and developed it into a car interior. Perhaps the most obvious brand reinterpretation was Maximillian Missoni's eCar, an imaginary collaboration with Ford, BP, and Apple, that exploits the computer company's instantly recognizable high-tech aesthetic (developed by product designer Jonathan Ive). The RCA's thinking is that, in the future, automotive brands will each represent a different ethos, rather than a simple association with a type of vehicle or price bracket. Understand the brand, and you can re-tool the product to satisfy consumer expectations.

Audi Avus Quattro:

The Audi Avus Quattro was shown at the 1991 Frankfurt Motor show, resplendent in polished aluminum bodywork. Heralding a sleeker, more sporting company image, the design preceded the more compact TT (first shown as a concept in 1995) as well as the use of aluminum in production car design.

Transport as Secondary Function

Mitsubishi's 2001 design competition illustrates a manufacturer's search for different interpretations of a single brand, in this case the future of the firm's renowned cross-country vehicle. The winning design, Yoshihiro Ishida's Personal Automotive Opportunity (PAO), was intended specifically for the Indian market. A "people's car" with intentionally elephantine styling cues, the PAO is trimmed throughout with rich local fabrics: it looks undeniably Indian, as Ishida intended, although, more practically, it is sizeable and cheap. The competition's environmental focus ranged from the bizarre – Akinori Myoui's symmetrical refleX, with matching panels to "help conserve energy and resources" – to the unlikely – Rik de Reuver's flimsy Quadcycle, about as far removed from the company's massive Shogun production cars as can be imagined. These mixtures of stereotypes and good intentions are work in progress, broad brush concepts that seem to bear little relation to the current state of the industry. But the point they raise – products should match the diversity inherent in a global market – is pertinent. Cars will become increasingly akin to products, seeking consumer trust and loyalty, and concepts are at the forefront of these new developments. If product designers can design cars, and respected car designers are talking about considering cars as just a consumer product, then the rationale informing the industry for half a century has altered: transportation is becoming a secondary function of cars. What is the future of the concept car?

Footnotes

i   Sex, Drink and Fast Cars, Stephen Bayley (Faber and Faber, 1986), p109

ii   Robert A. Lutz, GM, speaking at the Auto Week Design Forum, Detroit (9 January 2002)

iii See Audi's 2001 "Glider" TV commercial and 2002 "Jimi Hendrix" commercial, which stated that the guitarist's music directly influenced the TT's designers.

iv   Information Week magazine (3 June 2002)

v   www.ingenidesign.com

vi   'Grafting Hybrids: Designing the brand' by Sam Livingstone, Car Design magazine (2001)

Repackaging perfection

New Mini

The Mini ACV30:

This was BMW's first attempt at re-interpreting the Mini brand. Developed while the original vehicle was still in production, the rounded, stylized appearance was developed for the production version.

Sir Alec Issigonis' Mini proved so enduring that the iconic small car was eventually killed off by encroaching legislation, not lack of popularity. The Mini name passed to BMW when they acquired Rover, and the company cannily retained the brand when it sold the loss-making English company in 2000. Several Mini replacements had been proposed, from Issigonis's original 9X in the late 1960s to Rover's Mini Spiritual. The Spiritual was seen as a suitably radical concept with the same attitude of the original car. In contrast, when the BMW-designed new MINI first broke cover in 2000, it was criticised for being retrograde and revivalist, rather than a reinterpretation of the elements that made the original such a success.

Styled by Frank Stephenson, the key forms of the new MINI were first hinted at by the company's Mini ACV30 (Anniversary Concept Vehicle), developed at the same time as the Spiritual, only this time with BMW's involvement. The ACV30 was designed by Adrian van Hooydonk and built on MGF mechanicals. More curvaceous and sporting than the final car, it mixed Issigonis's original Mini with inspiration from the small sporting cars of the 1950s and 1960s, such as the Frogeye Sprite. The concept maintained the original's four-square stance, but softened the shape with accentuated wheelarches and headlamps. Importantly, the concept introduced the solid band of wraparound glazing that separates the waistline from the windows with a glazed-over B and C pillar. This "banding" is accentuated by the separately colored roof, harking back to the Mini's 1960s heyday. Both of these features were carried over to the production model. The concept was subsequently exhibited at MOMA New York's Different Roads: Automobiles for the Next Century exhibition in 1999.

Living legends:
Ford

A bonnet detail of the new GT40 concept, launched in May 2002:

Taller and longer than the 1960s original of the same name, the GT40 nonetheless is strikingly similar in appearance. A product of the company's Living Legends studio in Dearborn, Michigan, the new GT40 will be built in limited numbers from 2003, Ford's 100th anniversary year.

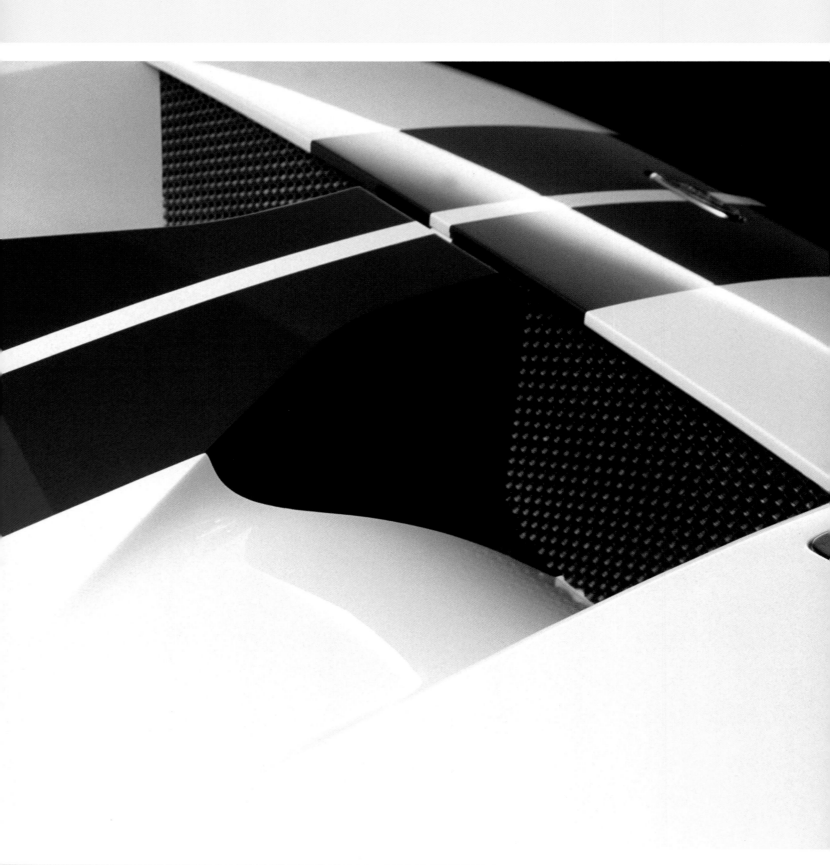

Living legends
Ford

Rear three-quarters view of the new GT40:

Even the graphics – the bold blue center stripe –
hark back to the 1960s original.

Headlight detail, GT40 concept:

The design integrates modern lighting technology
into the original's classic faired-in headlamp style.

Of all today's manufacturers, Ford have proved the most adept (and also the most controversial) at knowing when to look to the future or draw from the past. Like GM, Ford has a globally diffused brand portfolio, from the ubiquitous mass-market Ford brand, middle-market brands such as Mazda (Japan) and Mercury (US) through to the high-end marques of Lincoln in America, Land-Rover, Jaguar, and Aston Martin in the UK, and Volvo in Sweden. In 1999, Ford's European division launched the Focus, the production debut of the company's New Edge styling. New Edge hardened up the swooping curves that had dominated styling for nearly 15 years, reducing them to faceted planes that intersected with geometric precision. Although New Edge was a development of the 1970s origami school of car design – all razor-sharp edges and flat planes, it lent itself well to the Focus's one-box profile.

Remarkably, the Focus was accepted almost immediately, with the period known in the industry as the "soak time" being brief. This was primarily due to the gradual drip-feed of information about the design to the motoring public. The Focus replaced the Escort,

traditionally Ford's bestselling medium-sized car. Replacing this with a wilfully avant-garde design would have been suicidal if there was any chance the Focus would be rejected. New Edge styling debuted with Ford's GT90 concept in 1995, an updated version of the seminal 1966 GT40 race-car. The GT90 did away with curves almost completely, its flanks influenced instead by the faceted body of the Stealth fighter jet, complete with triangular lights, exhaust, and air intakes. While the GT90 was far removed from the Focus's sedate family car market, it nonetheless played a vital role in preparing and disseminating the new style.

The GT90 was a contemporary take on a classic car, but the company's subsequent decision to build a supercar based on the 2002 GT40 Concept was criticized. At first sight, the new car appears to be little more than a straight-faced update of the original. Following, as it did, on the decision to build a new Ford Thunderbird (after the 1999 concept that distilled elements of the classic 1955 and 1961 models into a contemporary idiom), industry watchers have suggested that the inward-looking approach is not a viable long-term strategy.

A passion for innovation

Alfa Romeo

1968 Alfa Romeo 33.2 Coupé Stradale:

This seminal road-going version of the Type 33 racing car was styled by Scaglione. Just 18 were made, with their wrapover gull-wing doors, huge faired-in headlamps and V8, race-tuned engine.

A passion for innovation
Alfa Romeo

Alfa Romeo's developing style:

(From top to bottom.) The 1938 Alfa Romeo Le Mans, designed by Vittorio Jano; the 1951 Alfa Romeo Gran Premio 159, a single seater race car with bodywork by Gioachino Colombo; the 1952 Alfa Romeo Disco Volante, an experimental sports car with bodywork by Touring; the Disco Volante formed the basis of the later BAT cars; the 1954 Alfa Romeo Giulietta Sprint, with bodywork by Bertone and the 1975 Alfa Romeo 33 TT 12, bodywork by Autodella.

The Alfa Romeo Brera:

Styled by the legendary Giorgetto Giugiaro, who founded
ItalDesign in the 1960s, the Brera was a highly practical
and eminently buildable concept car. Alfa Romeo decided
against proceeding with the design.

There are several different ways of dealing with concept cars,
explains Wolfgang Egger, Head of Alfa Romeo's Centro Stile in
Arese, Milan. One can either predict the future, create an essay
in technology, or undertake a pure design exercise. Straight away,
Egger excludes the third option. Speaking at the opening of
Sustaining Beauty, the company's 2001/2002 exhibition at London's
Science Museum, Egger sets out the fundamentals behind Alfa
Romeo's approach to the concept car: "Alfa is not about beauty
for beauty's sake – we must have technology and form together
in a concept."

The most influential Alfa concept to emerge in recent years was the
Nuvola, a low-slung sports coupé unveiled at the 1996 Paris Motor
Show and styled by Walter de Silva. The Nuvola concept, a fully
working vehicle (as are all of Alfa's concepts – "an Alfa without an
engine is unthinkable," according to Egger), is the stylistic source
for the marque's current range of saloons and coupés. The concept
coincided with the introduction of the slogan "Cuoro Sportivo"
("Sporting Heart"), and with it a new emphasis on key styling

elements. "We went back to the way we treated metal in the 1950s,"
says Egger, "and today's production possibilities are more
sophisticated." Styled in-house (although the company has often
used external studios, most notably Pininfarina and Bertone), Nuvola
was used as a means of finding Alfa's primal elements, summed up
by Egger with the five fundamentals: "beauty, strength, seduction,
expressiveness, and modernity" (the concept was the first car to
have LED brake lights, for example).

The Nuvola was preceded by the Proteo concept, a 1991 coupé
described at the time as, "a manifesto of the company's objectives
for the future," and an influence behind the subsequent Spyder.
Alfa's in-house team subsequently created the Centauri (1999), a
more radical departure. External interpretations of the Alfa tradition
include Pininfarina's Dardo (1998), the Bertone Bella (1999),
ItalDesign's low-slung Scighera (1997), and the Fioravanti Vola
(2001). All are faithful to the features that identify an Alfa Romeo,
most notably the heart-shaped radiator grill.

The concept as foretaste

Volkswagen

The 2001 Volkswagen Magellan concept:

Now entering production in a more conservative form, the Magellan
was a radical luxury off-roader, aimed at the adventure sports market.
An equipment trailer, removable Global Positioning System (GPS) device,
tough styling and a historically evocative name complete the package.

VW have spent the first years of the 21st century re-positioning their
brand upmarket and reorganizing the other marques in the VW
group to fill the gaps originally occupied by VW models. This has
seen the introduction of the record-breaking W12 supercar concept
(nearly, but not quite approved for production), the Phaeton
limousine, and the Tuareg SUV. The latter is derived from the
Magellan concept, introduced in 2001. The Magellan was the
company's first SUV, a compact manifestation of the profitable and
hugely popular genre that dominates the American market. Styled
in the increasingly popular "enhanced estate car" mold (like the Audi
All-Road, Volvo Cross Country, and Porsche Cayenne, with which it
shares some components), the Magellan was a great deal more
radical than the production model Tuareg. The concept came with
matching "utility trailer" and a roof-top luggage container that could
be flipped over and used as a small boat. Inside, chunky switchgear
and adetachable Global Positioning System (GPS) unit emphasized
the sports/action role of the vehicle.

Concept cars and corporate image

BMW

The four-wheel drive X-Coupe (2001):

This was the first hint of a new radicalism in BMW's design department, shunning the company's sober symmetry for a highly styled look. The X-Coupe, which has passed on genes to the new Z4, featured this unusual tailgate.

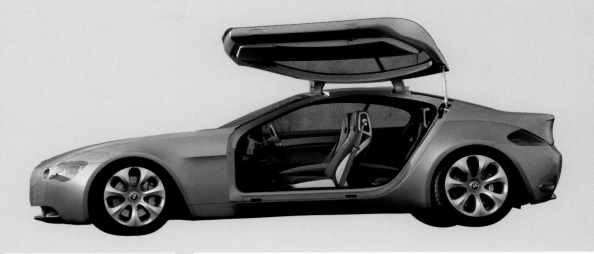

BMW's CS1 (right):

BMW's Z9 Gran Turismo (1999) (left):

The Gran Turismo was an influential concept that hailed the company's return to big coupés and debuted several new styling cues.

BMW
Concept cars and corporate image

Marketing the Dream

Few mainstream manufacturers have used concepts with such an explicit eye to the market as BMW. Their design has traditionally been gradual and evolutionary. The current 5-series, for example, has a visual history that can be traced back to the debut E12 model (BMW uses the "E" numbers as internal designations) in 1971, through the E28, E34, E39 and its replacement, due in 2003. Over a period of 30 years, the 5-series' essential qualities have remained constant, with incremental (and impressive) technical innovation concealed beneath relatively gradual bodywork changes.

All that changed with the arrival of Chris Bangle as BMW's Design Director in 1992. Through a series of increasingly radical concepts, Bangle paved the way for the introduction of the 2002 7-series, a bold departure from the corporate face that had evolved for so long. At Frankfurt in 1999, the company presented the Z9 Gran Turismo, a large, muscular two-door coupé that suggested itself as a replacement for the 8-Series. The car had unusual doors – ostensibly arranged in a gull-wing format, they actually included conventionally opening doors within the huge gull-wing frame, giving a choice of

ways to enter the vehicle. The arrangement of the bootline, with a profile raised above the line of the shoulder height, was carried through to the 2002 7-series, while the Z9 also saw the introduction of the iDrive information system, a single dial that controls access to all navigation, entertainment, and heating systems. Preceding the Z9 had been the Z8 roadster, originally presented as a concept but subsequently built in limited quantities.

The Z9's radicalism was followed by the X-Coupe, a product of the Designworks Studio in California. The asymmetric, four-wheel drive vehicle created controversy when it was first shown at Detroit in 2001. Of chief importance are the small car's "flame-traced" flanks, an innovative and sculptural way of treating the intersection between two surfaces, exaggerating the shape of the car's body. The X-Coupe will feed into the forthcoming Z3 and Z5 coupés, while the CS1 Concept Car, first shown at Geneva in 2002, is close to the 1-Series. The company hoped to sell one million cars for the first time in 2002.

The practical concept
Saab

Saab's 2001 9X concept:

The 9X streamlined the brand's core elements – most notably the radiator grille – while simultaneously debuting a type of car (small, flexibly configured sports coupé) that the Swedish manufacturer had long eschewed.

Saab is a sub-division of General Motors. Although mechanicals are shared with other GM brands, the Swedish firm retains a high degree of design independence. The company's most recent concept is the Saab 9-3X, which was almost immediately preceded by the 9X, Saab's first concept since the EV-1 coupé in the mid-1980s. This latter car, styled by Björn Envall, was perhaps more radical, entirely glazed above the waistline and incorporating solar panels to air-condition the passenger compartment.

In contrast, the 9X and 9-3X are more practical and production-ready. The cars are true cross-overs, compact sporting hybrids that combine four-wheel drive with a "shooting brake" appearance. The two-box design has an aggressive stance, reducing the signature headlight and grille arrangement into three stark shapes, and

carrying the shape of the wraparound windscreen back to the c-pillar. The glasshouse (or daylight openings – DLO) becomes visor-like, an effect heightened by the use of the concept's tinted windows. Unveiled at the 2002 Detroit Show, The 9-3X is visually similar to the 9X: both were styled by Saab's design director Michael Mauer, who has described the vehicles as a "road map" for the company's future direction, eschewing retro styling and emphasizing Saab's driver-focused approach.

Chapter Four:
Future

"Just as it has spearheaded progress in the past, the dream car will continue to search the future for automotive improvements symbolizing the limitless imagination of the stylists and engineers of the auto industry."

General Motors press release, 1965

On 16 June 2002 at Ford's Design Center in Dearborn, Michigan, some 51 design prototypes, including iconic and unique vehicles from the company's archives, were auctioned off by Christies. Prices for the concepts, few of which were street legal or even contained mechanical components, ranged from a few thousand dollars to $1.1m, indicating that concepts still have a remarkable hold on our imaginations. Ford was the first major manufacturer to sell concepts in this way, giving the cars a second chance at the limelight. The collection was eclectic, and included the influential Probe V aerodynamic study from 1985, the menacing Lincoln Sentinel from 1996, and numerous collaborations with Ghia, including the diminutive Trio from 1983, still the smallest four-wheeler to bear Ford's name. Some of the concepts had aged badly, their state-of-the-art technology rusted up and non-functioning. It was inevitable that these once futuristic visions should eventually appear outdated and tired.

Fifties Futurism
The futurism industry is constantly revising the shape of tomorrow and concept cars are a reflection of the changing cultural expectations of each succeeding generation. As the features and shapes demonstrated on concepts become part of the mainstream, we tend to forget their radical origins, choosing to move on and marvel at the next futurist vision. Will the current concept building system continue – will there always be a role for the concept car? Concept cars have always demonstrated things that were just out of reach; the self-parking car, the intelligent vehicle, the flying car. "Mark my words," Henry Ford was quoted as saying in 1940. "A combination airplane and motor car is coming. You may smile. But it will come." Ford's Flivver aeroplane – light, easy to fly and affordable – was designed with the backing of the American Bureau of Air Commerce, looking to revolutionize private transportation. Others went further: Waldo Waterman's Aeromobile (from 1934) and Molt Taylor's Aerocar (1959) blended the car with the plane, feeding the urban myth that America's interstate system was partly developed with a view toward accommodating future aerial traffic. The February 1951 cover of Popular Mechanics magazine shows a rain-coated commuter pushing his yellow personal helicopter into his garage, as his neighbour soars overhead[i], while in July 1957 a family was depicted in their 1967 model Hiller "aerial sedan." The post-war mania for concepts that, variously, flew, drove themselves, or were powered with strange new forms of energy (even atomic energy, with Ford's 1958 Nucleon proposal, which the company claimed would be able to travel 5,000 miles on a single charge) filtered down into the design of aviation-inspired production cars[ii].

2001 Peugeot Moonster:

Radical chic from a mass-market manufacturer, the Moonster is pure visual extravagance – with no possibility of manufacture. The concept is photographed with its designer, Marko Lukovic.

The Concept Matures

This futurism was naked and unashamed, space-age imagery with barely concealed overtones of jet age military equipment, rich with Cold War paranoia. Even when the industry sobered up, the dream car syndrome prevailed. The neat, rectilinear products of the 1960s and 1970s could be described as the motor industry's international style, homogenized, precise, and ultimately anonymous. Concepts were different. Show cars were fantasy objects far beyond the reach of the ordinary consumer, a form of escape from real world concerns such as the birth of the environmental movement or the Middle East fuel crisis. The 1980s saw the concept mature. A resurgence of interest in aerodynamics and streamlining, assisted by a better understanding of the science and the widespread use of the computer as a design tool, drove the search for the new forms of efficiency. California's controversial Zero Emission Vehicle mandate, finally introduced in 1990, was billed as the legislation of the future, demanding that 10 percent of new cars sold in the state in 2003 be electric or hybrid. Slippery, lightweight concepts proliferated, edging out the unbuildable, impractical supercars: the Buick Questor (1983), Saab EV-1 (1985), Ford's Probe Series (1979–1983), Pontiac Pursuit (1987), the seminal IAD Alien (1986), and GM's Impact (1990) all developed as a direct response to the Californian legislation. For a time, concepts rarely strayed into the realms of the truly bizarre. Worthy projects like Partnership for a New Generation of Vehicles (PNGV), DaimlerChrysler, Ford, and GM's proposals for an 80mpg vehicle that compared favorably to the average American car, failed to make a serious impact on a market more concerned with style than substance. If anything, the average concept began to look increasingly buildable as the dream car assumed the responsible role of design test bed. In the style and image-obsessed 1980s and 1990s, the conceptual automobile lost its political charge.

Avoidance Tactics

Paradoxically, during the dream car era, when concepts were steeped in politically incorrect presentations of speed, sexuality, and solitary thrills, there were heady debates about urbanism, mass transportation, and the ordering of cities[iii]. Architects, stylists, and manufacturers colluded, proposing conceptual mass transit systems that encompassed moving walkways, monorails, trolleybuses, and individual cars. Today, few contemporary concepts tackle the root problems of personal transportation, choosing instead to address problems of congestion and pollution with features that divert attention from traffic – turning a vehicle into a mobile office, for example, or making the vehicle resemble a piece of mobile architecture. We can no longer travel quickly and stylishly – we have to make productive use of our time stuck in personal transportation.

Buick Questor (1983):

The 1983 Buick Questor debuted in the company's 75th anniversary year as a rolling showcase of new technology, including voice-activated phone and laser door keys. The interior was angular, with matt black hi-tech styling.

A New Daring

Just as production technology and marketing saturation were bringing the concept car in line with the fashion industry's fickle attention, the truly inventive concept appears to be making a comeback. Naturally, there are raw commercial concerns behind this move, not least the fragmented state of the market and the lure of technological innovation to a new generation of gadget-hungry consumers. In the past five years, the most daring concept vehicles have been those to explore alternative power sources and new applications of information technology. GM's 2002 AUTOnomy is a case in point. Essentially a fuel cell-powered platform for future vehicles, AUTOnomy was intended to be "free from the confines of traditional automotive architecture." By starting with a blank slate, the vehicle was designed around fuel cell and drive-by-wire technology, doing away with the control systems and structures that have shaped the car for nearly a century. Developed by GM's Global Alternative Propulsion Center in Germany, the AUTOnomy's essential systems are contained with a universal, six-inch-thick "skateboard" chassis, allowing for different body styles to be developed entirely independently, then "docked" with the chassis.

As the avant-garde seeps into the mainstream, manufacturers must strive harder to appear truly radical. The 2001 Peugeot Moonster was a light-hearted concept vehicle born from a student design competition, aiming to look "outside the automobile sector for design inspiration." Winner Marko Lukovic, a comic strip and science fiction fan, created the Moonster, a computer render brought to life by Peugeot's Style Center, as a typical vehicle from 2020. This silvery, sculptural personal transportation module flies in the face of current futurist thinking about the world of 2020. At the end of the 1990s, steel manufacturers Corus developed a 2020 car in conjunction with product designers JAM. Their 2020 Vision took another approach, dividing the car into two components – the cabin and the "power cell." The former could be attached to either public or private transportation systems, or even as an extra room in an apartment building when not in use (reminiscent of the Lexus-developed concepts shown in the film Minority Report). These vehicles have an understandable self-promotional agenda: Corus, for example, want future vehicles to be built from steel, despite the need for lightweight, environmental efficiency. At the other end of the scale is the Giugiaro-designed Aston Martin Twenty-Twenty concept – a highly conventional extrapolation of current Aston Martin styling.

The evolving structure of the motor industry is an indicator of how car design will develop. Independent design studios function effectively as micro production centres, able to design, test, and manufacture a new car entirely in-house, the only difference being that technology facilitates swifter development times and cost-

General Motors Hy-wire concept 2002:

The Hy-wire is the most recent concept development of the fuel cell-powered car. Developed from the more obviously futuristic AUTOnomy concept, the fully functional Hy-wire was built at Bertone in Turin, pioneers of the drive-by-wire electronics that allow the car's clean, one-box lines and excellent interior space.

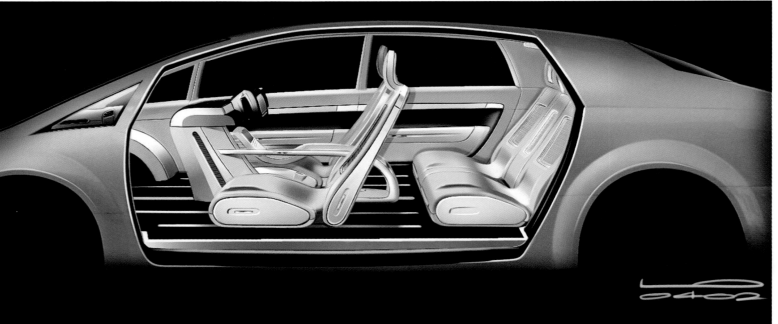

1999 Hummer H2:

The SUV has a powerful appeal despite its denial of the realities of congestion, pollution and depleted material resources. The H2 'domesticated' the Hummer military jeep, although its tough exterior remained the car's principal selling point.

effective lower production runs. ItalDesign employs 1,000 people on 12 sites, using computers to halve development times while increasing the number of proposals. Engineering and technical development co-exist with research into style and form. "As a result of the integration between the engineering digital mock-up phase and virtual testing, we can immediately check interferences, if any, and we can start producing the first prototype parts with technical and engineered control," explains Styling Director Fabrizio Giugiaro, son of Giorgio Giugiaro.[iv] ItalDesign was able to produce 33 styling models and five running prototypes between 1998 and 2001, including the Twenty-Twenty, which went from styling model to running concept in just two months.

Hybrid Designs

New cars are increasingly "bred" by mixing existing genres to generate new forms. Station wagons and estate cars have been superseded by the "sports tourer," a mix of coupé sleekness and load-carrying practicality; while the estate car has become entangled with the MPV to become the minivan, an enhanced version of the old-school load carrier. Crossover concepts go even further: Audi's Steppenwolf, the Fioravanti Yak, the BMW X-Coupé – all take established genres (hatchback, off-roader, and sporting coupé) and add a twist. This diversity means shorter production runs, with car companies fragmenting into smaller divisions able to target

consumer segments more accurately. At the 2002 New York show, Toyota USA launched SCION, a new brand aimed at the Internet-generation, who were "reaching driving age and entering the car market." As a result of research by the firm's Genesis Group, established in 1998 to study youth culture, two SCION concepts – the boxy bbX and the more radical ccX coupé – were launched, along with a web site[v] promising to "act as a two-way conduit between buyer and seller," encouraging brand loyalty and interaction. Crossovers are an acknowledgement that it's no longer possible to build a car that is all things to all people, as was once the case. The Ford Model T, the VW Beetle, the Mini were all cars that crossed cultures and classes. Now, Ford Ka, the MINI and the New Beetle, are all specifically targeted vehicles, cars which are almost caricatures of their former incarnations.

The Tonka Truck Syndrome

Consumer willingness to embrace novelty, regardless of practicality, will always exist: the burgeoning sports car market[vi] is based on desire, not need. The multi-functional object has always been a fallacy, an idea born out of our desire, rather than our needs. To a certain extent, this is reflected in the car market, where the primary characteristics of a vehicle are subsumed by our desire to use them in a different way. Thus, we have the Sports Utility Vehicle or SUV, a car designed as a capable off-roader yet rarely, if ever,

2001 Mazda Secret Hideout:

Retro-styled, with the emphasis on modest curves and a simple, almost functional interior, the Secret Hideout, as its name suggests, is a home from home for the younger generation.

used for that purpose. The SUV market uses concepts extensively to pump up the image of the vehicle that sits on the dealer's forecourt. The motor show podium is not a place for subtlety, and Ford and GM, together with the more outré Japanese manufacturers, strive to outdo each other in what has become a playground squabble on a global scale. Ford's mighty F-350 Tonka, GM's Terradyne (manufactured by GMC, the company's light truck division), the Hummer H2, Nissan's Pathfinder and Nails concepts – it's all too easy to just patronize these huge vehicles, denigrating the aspirations of those who hanker after this much size and power. Even J Mays was moved to suggest that Ford's bright yellow collaboration with Tonka would appeal to every American's inner child, blending the robust popularity of a hugely popular childhood toy with the physical security and environmental dominance afforded by its life-size progeny.

SUV concepts also deny the reality of congestion, pollution, and material resources. If, as many suggest, cities will ultimately re-organize themselves to accommodate a new breed of personal transportation – epitomized by Dean Kamen's scooter-like Segway – then surely the future of the car will be radically different? In the absence of affordable environmental technology, or the consumer demand for small cars, concept interiors (often created in totally separate studios) have become the new focus. Interiors prime the consumer for new materials, driving aids, technologies, and styles

as cars become ever more wired for communications and entertainment, allowing global concerns to be subsumed by a more comforting and immersive environment. An early example of the mobile internet, BMW's TX5 was a rolling demo vehicle stuffed full of communications technology. Based around the company's X5 SUV and developed by the firm's Technology Office, the TX5 is a taster of the information-rich roads of the future, when futurists envisage cars and other vehicles having their own IP addresses, becoming additional nodes on the global communications network.

Anthropomorphism

Information technology is also creating concepts that develop our relationship with our cars, perhaps one of the most anthropomorphized and fetishized relationships we have with an inanimate object. Toyota claim their Pod (see page 152) "explores the potential for communication between people and their vehicle," a communication which hitherto has been almost entirely one-way. The inherent ability of the human brain to extrapolate and enhance familiar patterns – such as the features of the face – from an arrangement of objects, is exemplified in the Pod. Other concepts also seek to present a cute, natural face, with headlights and radiator grille smiling or snarling.

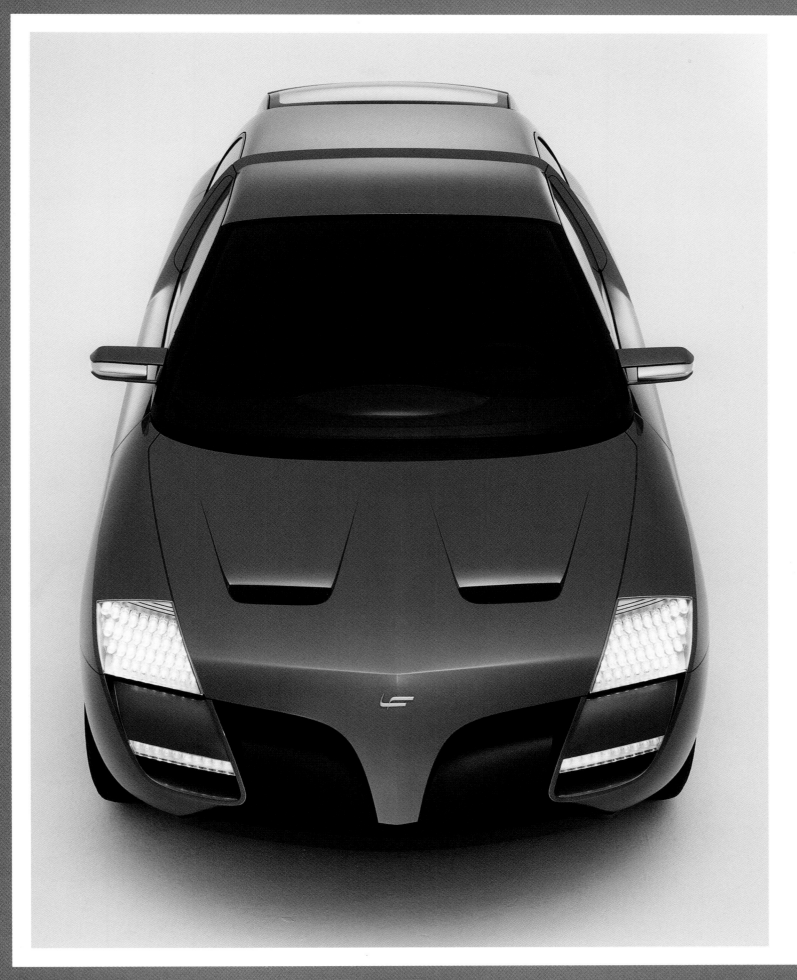

2002 Fioravanti Yak:

This small studio, started by ex-Pininfarina managing director Leonardo Fioravanti, has created stunning concepts based on Ferraris and Alfa Romeos. The Yak was the company's take on the crossover 4x4, a V8-powered machine that incorporates side window wipers, a roomy cabin and high levels of performance.

2001 Opel Frogster:

Opel's undeniably cute Frogster concept, shown at the 2001
Frankfurt show, is likely to be built. With a name that harks
back to Opel's early days (in this case a 1920s model called
the Laubfrosch, or "tree frog"), the concept demonstrated an
all-new feature – a hard-wearing metal cover that rolled back
to protect the interior when the car is parked. Whether this
will translate into a production feature isn't known.

2001 Nissan Chappo:

Demonstrated at the 2001 Geneva Show, the clean,
white lines of the Chappo were almost anti-style –
this was a car for being in, not looking at.

Interior Spaces

With this technological playground comes a desire to make new space within the city. The 2001 Tokyo Show saw a plethora of concepts that addressed internal space, not external appearance. Mazda's Secret Hideout was perhaps the most blatant, a contemporary update of the customised vans favored by global teenagers worldwide and intended as a private space away from the cramped Japanese home. Nissan's Ideo concept placed the emphasis on communications technology, a "net vehicle" that processes a constant stream of environmental information on its dashboard. The company's Kino (German for cinema) is a more achievable MPV, complete with comprehensive entertainment options. The Chappo, shown earlier in the year, was another "outside-in" design – a social space, not just a car. Integrated alternative transport systems were also common – witness the bikes stowed in Honda's Unibox (see page 148) and Bulldog concepts. Isuzu's Zen (see page 94) also made deliberately architectural references.

The trend for single box, interior-focused concept vehicles has continued. The 2002 Pforzheim Design Forum in Germany exhibited a marked emphasis on architectural, almost static forms. The forum, which brought together 12 students from six colleges around the world, was based around creating a new vehicle that would be uniquely suited to the students' country of origin. In particular, Koh Takaya's Nagomi Peace concept, demonstrated how design cycles are returning full circle, re-emphasizing almost carriage-like symmetry, and containing Japanese architectural features such as the tatami mat. "Design is about getting under the skin of things," says Dale Harrow of the RCA, pointing to the explosion of interest in "touch, feel, and texture," the tactile means of conveying quality. As well as post-graduate students, the RCA has several PhD research fellows examining new vehicle technologies and emerging typologies, such as the compact city car.

The Japanese City Car

The small, dedicated city car offers clues to the future of the car, concept or otherwise. Japanese firms, experienced with their country's high taxes, cramped urban confines, and relatively rigid cultural mores, have traditionally been more experimental in their approach to tomorrow's automotive forms. Legal incentives favor microcars – K-cars, or *Keijidousha* – and have stimulated the genre, encouraging manufacturers to experiment with body styles within the restrictions on engine size and dimensions (14.4ft x 4.9ft, with a 660cc engine). The 1995 Daihatsu Town Cube was a pure rectangle on wheels, exactly filling the micro car's dimensional constraints. While the Town Cube concept was tongue-in-cheek, maximizing space while retaining individuality and character remains high on manufacturers' agendas. Packaging and styling can run riot, with sporting, retro, futuristic, and eminently practical designs crowding the microcar sector.

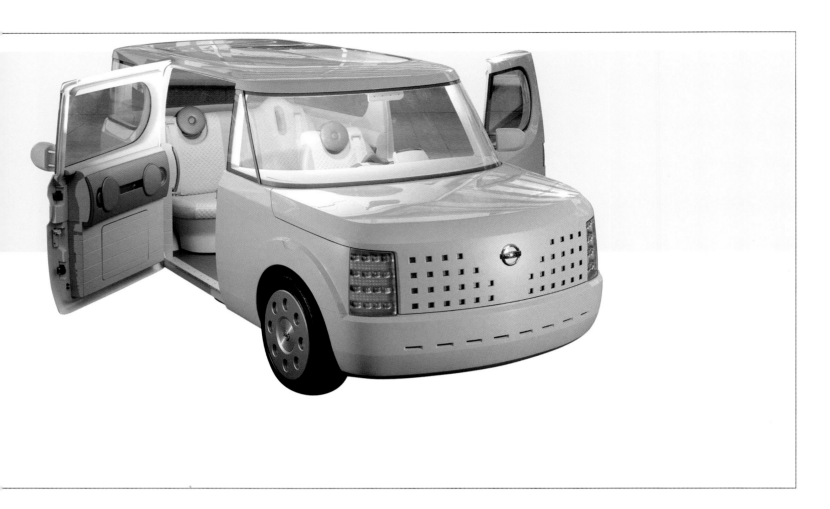

American Retro

If the Tokyo 2001 show was the apotheosis of in-car technology, then Detroit 2002 looked like a return to America's glory days. Design chiefs often describe brands as nameplates, and the American industry has more evocative names than most – Thunderbird, Mustang, Camaro, Charger. The retro concepts that dominated the show – from Ford and GM in particular – saw the past being plundered – all long, low lines, excessive width, and even hints of fins and chrome. The classic 1950s auto was the product of planned obsolescence, with deliberately exaggerated physical characteristics used as symbols of quality and prestige. Models were advertised as being "longer, lower, and sleeker," year on year. Although today's heavyweights are by no means a return to the earlier era's technological short-comings, they are instead part of America's mission to revive its own self-esteem. A country that filleted its own manufacturing industry as economic growth shifted to consumer spending now finds itself in need of industrial icons. These concepts are Detroit muscle cars for the post-baby-boom generation, a cup-holder culture that needs all the contemporary comforts it can get.

Predictions

Just 20 years ago, and features like information technology, sustainable power plants, satellite navigation, sonar parking sensors, and materials recycling were the stuff of science fiction. In fact, so alien were some of these technologies – the mobile internet, for example, that no concept had the audacity to predict them. In-car telephones were the heights of communications technology. Conversely, the conceptual and stylistic extremes of earlier eras now appear laughable – the obsession with flight, automated roads, and sliding, folding, easy-access interiors. Today's concept car rarely matches this era of blue sky thinking. We are lulled into a passive acceptance of technology's relentless march: consumers almost believe that a technology exists long before it's even been designed. It's not a question of if, but when. Many of the concepts shown at major motor shows are far from innovative. "Let's not forget that the cars designed by and for the major manufacturers are sales boosters," points out analyst Robert Cumberford[vii], adding that, "In the majority of cases, concept cars are nothing but makeovers of production models finished long before they go into production, so that there's nothing new about them, new though they may be to the public." Cumberford calls show cars "entertainments," suggesting that they are simply styling exercises, far removed from the "conceptual watersheds" of the past.[viii] In the 12-month period to mid-2002, approximately 130 new concept and production vehicles were produced by the world's major car manufacturers.

Concept cars creep into the corners of the consumer psyche, making fresh associations, triggering new desires and wants.

The ultimate aim is to sell cars, so it's not surprising that colleges and design schools find themselves prioritizing marketing – the car stylist of the future will need to know their brands as well as their broad brush techniques. Perhaps one day the multi-platform market, an octopus of manufacturing allegiances and technology sharing, will make the concept obsolete altogether.

BMW's Technology Office, amongst others, is investigating rapid manufacturing, a development of the rapid prototyping technology already operating within the industry. Manufacturers will soon be able to produce complex variants of production models. Today's sophisticated production lines already work on an order-only basis; cars built to specific customer requirements – say green paint with red leather trim and a CD player. This combination of options doesn't exist at the time of the order, so it is added to the production queue at the factory. Simple choices at first, but as the options become more varied, rapid manufacturing – especially that which allows production line tools to be swiftly made up – will increase. CAD has irrevocably changed the way cars are designed and tested; it might one day change the way we buy our car.

Unlike in product design, where styling and configuration concepts are discrete, inevitably non-functional, and usually far from the public eye, the concept car has grown up in public, taking a central role in

the motoring industry. Optimistic futurists are looking to technology to radically enhance car design. Science fiction materials such as memory metal (panels predefined to a particular shape during manufacture, enabling them to return to their correct shape after a light accident, for example), structural glass, or the fuel cell platform developed by GM, will set stylists free from existing conventions.

European and Japanese motor manufacturers are increasingly led by design, rather than engineering, as manufacturing reaches a level of sophistication that makes purely technical distinctions between cars irrelevant. In America, however, the fantasy emphasis remains. Perhaps 60 percent of cars sold in the States are SUVs, a vehicle type that has little in common with real world needs. The RCA's Dale Harrow believes that having a restrictive, straightforward brand identity makes it more difficult for a company to be innovative. European companies tend to be more straightjacketed, tied by decades of tradition and a carefully constructed image, bolstered by marketing, and reinforced by the concept cars that appear at the major motor shows as flagship styling exercises. In contrast, Asian companies place more emphasis on innovation, using concepts in an unceasing quest for novelty that incorporates changing cultural values and attitudes towards youth, transport, technology, and the city. The increase in design and manufacturing speed has made car design more immediate, relevant, and influential.

1999 Daihatsu Copen:

Daihatsu's Copen, announced in summer 2002, was directly descended from the show car of the same name exhibited three years earlier. Remarkably, the production Copen showed no sign of toning down the lines of the concept two-seater shown three years previously at Tokyo: quite the opposite, as the production Copen was more explicitly conceptual in terms of its straightforward, two-seat configuration and emphasis on motoring as entertainment. Other concepts from the Japanese company demonstrate a similar verve – like the U4B, an urban 4x4 buggy, or the FF Ultra Space concept, and Muse microcar.

Automotive technology will continue to evolve – and there will always be a need for full-size concept models. The concepts that are presented to the public are as free from compromise as possible, yet are hardly the purest forms of design. In essence, they are distractions and deceits, cunningly worked up from materials alien to the production line. They exist because we demand an insight into the future, a reassurance that tomorrow will be better and brighter than today. The concept system will continue, just as long as there is a demand for new cars. Stylistic dead-ends and technological failures are swiftly forgotten as new concepts come year after year: it's up to the coming generation of students, designers, stylists, and design chiefs to exploit this new immediacy. After all, experience has taught that the unexpected path is often the one that triumphs.

Footnotes

i   Atomic Age: Art and Design of the Fifties, Marc Arceneaux (Troubador Press, 1975) contains discussion of the more elaborate concepts from the era.
ii   Paul Moller's Skycar also remains a concept, albeit one that the Californian inventor is determined to build – allegedly spending $70m on development. www.moller.com
iii New Movements in Cities, Brian Richards (Studio Vista, 1966).
iv   Fabrizio Giugiaro, Styling Director at ItalDesign, in a speech to Autoweek Design Forum, Detroit (9 January 2002)
v   www.scion.com
vi   SUV sales were predicted to double from around 1m units in 1990 to just over 2m units by 2001. In the event, 3.5m SUVs were sold in 2001 and sales continue to increase. The SUV is the most profitable market segment for car manufacturers.
vii 'Opinion column' by Robert Cumberford, Auto & Design magazine no. 132, p.10
viii The Car Design Yearbook 1, Stephen Newbury (Merrell, 2002)

GT Concept:

Just as Gran Turismo, the PlayStation racing game developed by the Japanese developer Polyphony Digital, introduced Western audiences to a whole generation of Japanese supercars (often at the expense of their European equivalents, thanks to manufacturers' reticence to licence their vehicles for the game), so the stand-alone sequels Gran Turismo Concept 2001 Tokyo and Concept 2001 Geneva have enhanced the image of concept cars – turning the fibreglass shells into virtual models that can actually be driven.

The personal concept car
Xeno

Nick Pugh's Xeno III is described by its creator as a "personal concept car," a bespoke vehicle designed to a precise specification. The Xeno III is a prototype for this new way of thinking, and the faceted, skeletal design allows for modular improvements in both technology and materials.

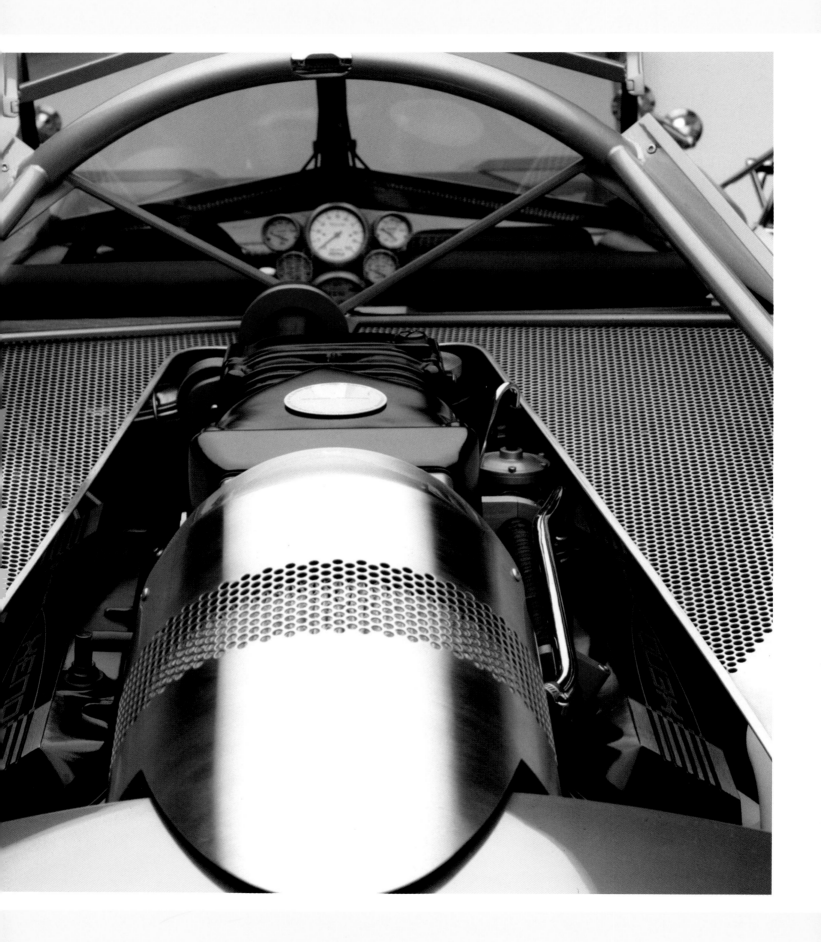

Future

The personal concept car
Xeno

Side view of the Xeno III two-seater:

The car wraps itself around the driver and passenger.
Details – lights, indicators, instruments – are highly individual. The car
was initially designed to take a closed loop fuel cell propulsion system.

Created by designer Nick Pugh, the Xeno III is styled as a personal concept car. Having studied at Pasadena's Art Center College of Design, Pugh has spent the last 12 years working on this singular vision. The Xeno III is a rich, gothic fantasy, drawing inspiration from science fiction and a contemporary extrapolation of Victorian-era engineering. A low-slung, natural gas-powered sports car, the Xeno III concept is skeletal and faceted, a strange combination of the hand-wrought and machine-styled. "In 50–100 years I see vehicle architecture becoming a true extension of human anatomy" Pugh says, demonstrating how structural elements are combined with the passenger compartment, effectively melding the driver into the design.

For now, only the very rich can afford to construct automotive dreams such as this (the Xeno was advertised for sale at $1m), but Pugh envisages a future when consumers will be able to bring their own data into a product's manufacturing process, customizing the end result to their precise specifications. He calls this "personal product manufacturing," foreseeing that as production technologies improve,

advanced rapid prototyping machines will evolve into just another domestic appliance; homes will have their own manufacturing plants.

The flexibility envisaged by this new era of manufacturing isn't a means to an end. "I think there will always be a place for generic mass-produced design," says Pugh. "The personal concept car is a special idea, but it will always make sense for a large proportion of vehicles to be the same as each other." He stresses that the Xeno is a "concept car of an idea, rather than a concept car that represents a model or trend." In this sense, the vehicle's radical appearance is a distraction, an indication of what the product of his envisaged personal manufacturing process would be if he was in charge. With the motto "a unique car design for every unique individual," the personal concept car is perhaps a natural extrapolation of the contemporary obsession with concepts, automotive or otherwise. "I think it is inevitable that the big companies will offer cars this way when the technology turns production into a direct digital-to-digital process, eliminating hand fabrication and assembly." www.nickpugh.com

Honda

The future according to Honda

The Unibox:

Truly conceptual and hugely radical, the Unibox is a piece of mobile architecture. Traditional body construction has been replaced by translucent panels attached to a visible structural frame. External airbags protect pedestrians in cramped urban environments.

Honda's Unibox is less car, more building; taking elements from both architecture and public transport. This radically "no-style" vehicle has a city bike integrated into the door, along with two small personal scooter devices. The exterior is clad in transparent polycarbonate panels, attached to a structural aluminum frame in much the same way as a contemporary building. The wheels – one axle at the front, two at the rear – are down-sized to increase internal space, while the interior is presented as modular and flexible, with the emphasis on storage, space, and practicality. This integration of architecture and transportation is looking many years, perhaps decades into the future, a sign of a confident manufacturer keen to predict and shape possible societal shifts in the use of transportation. The Unibox is a true concept, in the sense that its values and aesthetic are as alien to this era as the turbine-powered Firebirds were to the 1950s.

Secure innovations
Volvo

Volvo Safety Concept Car (SCC):

Designed for high visibility, both for drivers and other road users, the SCC is bright orange. A semi-transparent A-pillar improves the view of the road ahead and the car is stuffed full of life-saving and driver-assisting technology.

Part of Ford's brand portfolio, Volvo's reinvention as a curvier, less visibly austere manufacturer slightly compromised the company's reputation as being fastidiously committed to safety (ironically, although Volvos became sportier and more desirable, they also became safer). The Safety Concept Car (SCC) was designed to showcase how safety features could be integrated into an utterly contemporary shape. Described by the manufacturer as representing, "70 years of Volvo innovation rolled into one," features include lights which adapt to the curve of the road (first seen on Citroën's DS, but considerably enhanced), a night vision system, transparent A and B-pillars to improve forward visibility, intelligent brake lights, and rear-view cameras. Resplendent in high-visibility orange, the SCC is tangibly reminiscent of the Volvo 1800ES. Very unlikely to be manufactured in its current form, the concept was intended as a rolling demonstration of forthcoming safety technologies.

www.conceptlabvolvo.com

The concept as friendly technology

Toyota

The Toyota Pod:

Cheerily styled and aimed squarely at a networked, media-rich future, the Toyota Pod takes anthropomorphic design to extremes, reflecting the driver's mood.

Toyota's Pod was one of several futuristic/IT crossover concept vehicles shown at Tokyo 2001 by Japanese manufacturers. The Pod melded the national obsession with robots, crystallized in 2000 by the launch of Sony's Robot pet, the AIBO, and the ongoing exploration into automotive IT. The car is perhaps the most anthropomorphized consumer object, complete with smiling (or snarling) face. Designed in conjunction with Sony, the Pod even has a tail – or radio antenna – which can be wagged to create further expression. Styled like a little robotic insect, in keeping with the Japanese cartoon tradition, the Pod exaggerates its anthropomorphic qualities, with color-changing lights in its face that express different moods. These moods are partly derived from the driver – sensors integrated into the steering system diagnose how the driver is feeling and adapt accordingly, playing calming music, for example.

The concept as technical testbed

Bertone

(From left to right): BAT 5 (1953), BAT 7 (1954), BAT 9 (1955):

The celebrated Alfa Romeo Berlina Aerodinamica Technica cars (BAT) were
developed by Nuccio Bertone on standard Alfa 1900 Sprint chassis. Intended
as styling exercises for the replacement for the 1952 Disco Volante, the highly
aerodynamic trio of BAT cars were unveiled at the Turin Auto show.

Overseen by the design genius Nuccio Bertone (1914–1997),
Bertone is one of the great Italian studios, which has seen the concept
transformed from dream car to motor show icon. Like many studios,
the Turin-based firm began producing limited model runs on
separately supplied chassis, eventually graduating to designing
bodywork for the major manufacturers. In the 1950s, the company
produced a triumvirate of extraordinary aerodynamic concepts for
Alfa Romeo. The Berlinetta Aerodinamica Technica (BAT) series
pushed the body style of the period to extremes, with great curved
fins that almost touched at the tail. These vehicles were perhaps the
most radical European manifestation of the dream car aesthetic.
Today, Bertone has become a specialist manufacturer, best-known for
its cabriolet expertise – building models for Fiat and Vauxhall,
amongst others – and other specialist vehicles like BMW's C1
motorbike.

Concept cars have always played a major role in the company's
output, with commissions from a wide variety of manufacturers,
including Citroën (the Zabrus coupé of 1986), Oldsmobile
(the 04 roadster aimed at attracting younger buyers to the
brand), and Lamborghini (the 1988 Genesis, essentially a Multi-
Purpose Vehicle [MPV] powered by a V12 Lamborghini engine).
Recent vehicles include the Filo, Slim, and Novanta, the latter created
to celebrate the firm's 90th birthday. Filo and Novanta are both
technological as well as stylistic showcases, featuring drive-by-wire
technology adapted from the aerospace industry, which in turn
allowed the car's interior to be radically re-thought, without the
need for conventional pedals. The Slim was a city car that sat two
people in tandem configuration.

D1364464

Acknowledgements:

The author would like to acknowledge the invaluable support of Alex Adie and Mark Guard, not to mention the help from the many designers, studios, press officers and manufacturers who provided material and information for this book.